Bacteria

Bacteria

The Benign, the Bad, and the Beautiful

by

Trudy M. Wassenaar

Illustrations

Karoly Farkas

With photographs by

Eshel Ben Jacob

A John Wiley & Sons, Inc., Publication

Published by John Wiley & Sons, Inc., Hoboken, New Jersey
Published simultaneously in Canada

Library of Congress Cataloging-in-Publication Data:

Wassenaar, Trudy M.
Bacteria : the benign, the bad, and the beautiful / Trudy M. Wassenaar.
p. cm.
Includes index.
ISBN 978-1-118-10766-9 (hardback)
1. Bacteria. 2. Bacteria–Ecology. 3. Microbiology. I. Title.
QR74.8.W37 2012
616.9′201–dc23

2011021003

Printed in the United States of America

oBook ISBN: 978-1-118-14339-1
ePDF ISBN: 978-1-118-14336-0
ePub ISBN: 978-1-118-14338-4
eMOBI ISBN: 978-1-118-14337-7

10 9 8 7 6 5 4 3 2 1

Contents

Preface

This book is meant to be an antidote.

An antidote to neutralize some of the bad press that bacteria frequently receive, which could easily leave the impression that bacteria's main aim is to attack us, make us sick, or even kill us. Such press imprints us with the idea that all bacteria are mostly concerned with one goal: to damage humans where and when they can. This does not apply to the vast majority. Despite the few nasty ones out there, most bacteria are completely harmless, and this book tries to restore their reputation.

This book is an antidote to balance all existing books on biological subjects that completely ignore bacteria, pretending they are not important. As if a plant or animal could survive without bacteria (they cannot). As if mammals do not carry more bacteria than body cells. As if these bacteria living inside and on us do not carry more genes than our own genome does. Let's face it: we are a living support for bacteria, and these little wonders of evolution made us what we are.

The antidote is meant for future students who will likely be taught only a fraction of the knowledge that is available about bacteria. Medical microbiology mainly teaches about bacteria that cause disease, ignoring the fact that these account for a small minority that inhabit the bacterial world. Food microbiologists may concentrate on food poisoning, bacteria needed for food processing, or even those causing spoilage, but they often remain oblivious to the bacteria that helped these plant and animal foods to grow in the first place. Marine microbiologists learn little about bacteria living in soil, and so on. Why not start with a little knowledge about everything, before specializing. It will stimulate an open mind to science as a whole.

This book is also meant to be an antidote for all those who think bacteria are boring. Obviously they are not; otherwise, these pages could not have been filled.

However, the fact that you intend to read this book means that you display an interest in bacteria, or at least you are willing to make their acquaintance. You may thus not be the audience for which the antidote would be most effective, but at least it is hoped your interest is rewarded, and that you enjoy reading about this invisible subject, bacteria, which can be bad, benign or beautiful, but certainly not boring.

Note to the Reader

No prior knowledge about bacteria by the reader is assumed, although a general acceptance that bacteria exist, and some rudimentary knowledge of biology and chemistry is expected (and a faint idea about medicine, geology, geography and history). All scientific concepts are explained the first time they are introduced but used without much explanation in following chapters. Although all chapters can be read independently, it is, therefore, better to start with the first chapter and then just keep reading. The use of jargon and technical language is avoided, with the exception of scientific bacterial names. A glossary of scientific terms at the end of the book may prove helpful.

All bacterial species that feature in this book are listed in the alphabetical index at the end. A subject index is also provided.

Students who are interested in in-depth information might find the bibliography useful, where scientific literature is listed for each chapter.

The interested reader might be tempted to investigate particular subjects further, for which the Internet is a great resource. Such initiatives can only be encouraged. However, a considerable proportion of the information on microbiology provided by this medium is imprecise or conflictive, if not completely wrong. All subjects treated in this book were carefully checked with recent scientific literature, and a current consensus of opinion is presented where possible. Scientific interpretations sometimes differ between sources, and in these cases the choice of presented options is subjective.

The characters of Mrs. White, Joe, and Liz are imaginary. Scientific and noble titles of existing persons have been omitted throughout.

1

The Blue Planet

Imagine an intelligent species of extraterrestrial life that decides to explore our planet by sending sensors that report back visual information. Since these living beings are intelligent, they know that they should not base their judgments on single observations, so they send five sensors to different locations. These sensors happen to land in a city, in the ocean, in a forest, in a dry spot of a desert, and on Antarctica. They are equipped with very sensitive lenses that can see details at a micrometer scale, a thousandth of a millimeter. What would they report back to their home planet?

The extraterrestrial scientists would not bother about the single observations of humans and buildings reported by the first sensor. Nor would they pay attention to the fish and other marine animals, to the trees and plants, rocks and sand, or snow and ice, reported by the other respective sensors. These all appear to be location-dependent phenomena. The common and consistently reported observations would indicate the presence of bacteria, which are detected by each sensor, in large numbers and with great variety. These tiny cells appear omnipresent. The scientists would report, in their extraterrestrial newspaper, that the blue planet is the home of bacteria.

Bacteria: The Benign, the Bad, and the Beautiful, First Edition. Trudy M. Wassenaar.

For a long time in history, bacteria were indeed the only living things on earth. Our planet is 4760 million years old, and approximately 1100 million years after its formation, bacteria already existed. We know this from fossil stromatolites, mushroom-shaped formations built of layer after layer of bacteria, that usually grow in shallow waters of coastal areas. These look strikingly similar to the stromatolites that still grow today, and are produced by Cyanobacteria. The fossil stromatolites are 3.6 million years old and can only have been formed with the help of bacteria that lived in those early times. Other than these scarce remains, we have little information about early life on earth. The marine sediments and continental rocks where bacteria may have been deposited have been flattened and heated over time, or recrystalize completely, so that few rocks (and fossils) of this age have remained in their original shape. Moreover, bacteria do not easily leave fossil prints, as they lack hard body parts.

It took at least another 1500 million years before the first *eukaryotes* evolved to which all higher organisms belong. These have larger cells than bacteria, and their cells contain compartments, called organelles, that are specialized for particular functions. Eukaryotic cells have their DNA stored in a nucleus, in contrast to bacteria, which do not possess a nucleus, but rather have their DNA floating as a condensed structure in their cell soup. Bacteria belong to the so-called *prokaryotes*. There are more differences between prokaryotes and eukaryotes. Most cells of eukaryotes but none of the prokaryotes contain organelles dedicated to energy production, called *mitochondria*. Their cells further contain architectural structures that bacterial cells miss. The name eukaryote is derived from "good (or true) kernel," emphasizing the presence of a nucleus. The first eukaryotes that lived on earth were single cells, just like the unicellular eukaryotes that still exist today. The exact time point when these first eukaryotes arrived is as uncertain as that of the first prokaryotes, but estimates vary between 1900 and 2100 million years ago.

The next important step in the evolution of life was the formation of multicellular organisms, in which different cells perform different functions. This is the evolutionary line to which animals and plants belong. Multicellular organisms are all eukaryotes, and it is thought that life forms based on multiple cells evolved a number of times, independently, whereas only one of these evolutionary attempts led to the plants and animals we know today. Recently, thumb-sized fossils that are the remains of multicellular organisms that lived 2100 million years ago were discovered in Gabon (Africa). But it is uncertain whether they are the remains of multicellular eukaryotes or of large and complex colonies of bacteria. Other findings that report the existence of multicellular organisms 1600 million years ago have also failed to answer this important question unambiguously. Fossilized multicellular eukaryotes that are undisputed are only 600 million years old, by which time clearly plants and animals are recognizable.

The difference between a true multicellular organism and a colony of multiple bacterial cells is that in the former, cells perform different functions depending on their location within the organism. Moreover, every individual (of the same age) of a multicellular species is equally shaped, and the function of each cell can be predicted from its location within the organism. Finally, their individual cells cannot survive on their own, but only as part of the organism. Bacterial colonies can also contain specialized cells, and their function may even be location dependent, but the shapes of these colonies are not strictly conserved, and their cells can also survive independently.

The development of early life on earth is still riddled with uncertainties. There is no discussion, however, that bacteria existed before eukaryotes, since eukaryotes evolved from bacteria. Bacteria ruled for about half of the 3600 million years that life exists on earth and seem to have been alone for the first 1500 million years of their existence. Had the sensors from extraterrestrials arrived earlier, they would still have reported the dominance of bacteria, should they have stumbled on life at all.

Every surface around us is covered with bacteria. One drop of seawater contains a minimum of one million bacterial cells, at least up to a depth of 200 meter; the number decreases by a factor of ten as one dives deeper. Bacteria live in all depths of the water column, all the way down to the bottom of the ocean. One gram of soil also contains roughly 10 million bacterial cells, of types different from those found in seawater. Deeper soil will hold fewer bacteria, but no surface on earth is naturally sterile, apart from places where temperatures soar, such as volcanoes. Add to the marine and soil bacteria those living in submerged sediments, another major habitat full of life, and one can estimate the total number of all bacteria on earth at a given time. It results in a number written with 30 digits, give or take a digit. That is 100 million times the number of stars in the universe. These bacteria have mass, and although each cell is minute, together they add up to a considerable amount of biomass, nearly as much as that of all plants and animals together. This means that roughly half of all biomass on earth is invisible to the naked eye.

Bacteria are not only plentiful but also belong to many different types. Bacterial species outnumber any other biological life form in diversity, especially if we take the unknowns into account: it is estimated that we have characterized only one percent of all bacterial species that exist. Obviously, organisms invisible to the eye receive less attention than living things that we see around us, and doing research is hard if you can neither see nor grow the bugs. Most bacteria are not able to multiply under conditions that we can provide in a laboratory. Nevertheless, since we can isolate their DNA, we can still estimate how many of the "unknowns" exist. It is possible to determine whether DNA is of bacterial origin (as opposed to DNA from eukaryotes), and if so, whether it belongs to a species, or group of species, that has already been cultured and described. From such explorative DNA studies, we know that 99% of all bacterial species living in the oceans, in soils, or in sediments have never been cultured.

If only we could see them without the necessity of a microscope! How useful it would be if we could see the bacteria lurking on the glove of the surgeon who, after meticulously scrubbing his hands and underarms, unknowingly touched his hair with his gloves—bacteria that will smear onto the scalpel; enter the wound as the metal cuts through the flesh; start multiplying in this new warm, moist, and nutrient-rich environment; and cause a postoperative infection that sets back the patient's health by days or weeks.

How practical it would be to be able to see the bacteria growing in a freshly made desert that had raw egg whites and sugar as key ingredients and is now standing on the kitchen bench to set. One egg contained *Salmonella* bacteria, which feed on the sugar and start multiplying at the ambient temperature to reach critical numbers in only a few hours, and which will make half of the dinner party guests go down with diarrhea. Or to see the slightly bent bacteria that wiggle in the foul drinking water of a refugee camp, water that got contaminated by a single child suffering from the onset of diarrhea, which rapidly developed into cholera that will take hundreds of lives in the epidemic to follow. Imagine we could check the bacteria that were added to a health drink because they are so good for you, at least according to the commercial, but that in fact are all dead, and possibly useless, by the time the product was bought. What if the farmer could see the bacteria that start to attack his crop, way before the leaves are getting spotty and pale, a sure sign that the harvest will not be good this year.

By being able to see the bacteria that are ever-present, we might learn to tell them apart. Furry bacteria covered with thin hairs, bald ones, bacteria that hide themselves in a slimy layer of gel, or that bear a sturdy armor of protein for protection. Long bacteria and short and stubby ones, round or rod-shaped bodies, spiral or crooked cells. Communities of bacteria that grow into threads, bundles, heaps, or layers of cells. We could learn to recognize the dangerous ones, to avoid or kill them, and to leave the others alone. No longer would we need the soaps, cleansers, toothbrushes, dishwashing or laundry detergents, hand lotions, and all the other products that the advertisements make us believe would be unsafe to use without the addition of antibacterial agents. Being able to see and recognize bacteria would tell us that, in most cases, such additions are completely unnecessary, even ineffective, or useless.

Consider a commercial claim of a product killing "ninety nine percent of all bacteria in your sink." That would mean that, of the ten thousand bacteria typically living on a square inch of the wet sink surface, one hundred would survive. These have only to multiply for eight generations to be back to the numbers they were before the detergent was used, which would usually take them a couple of hours, provided there was enough food for them. To remove their food by cleansing with a detergent lacking antibacterials would have been just as efficient. And quite possibly, none of these bacteria would be likely to cause disease, so you might as well leave them alone. In fact, antibacterial household products do a lot of harm, as they selectively allow growth of bacteria that are resistant to their toxic components, and once these resistant forms dominate, they are more difficult to eradicate with antibiotics in case of emergency, for instance during an acute and serious infection.

Leaving bacteria alone in your kitchen sink probably goes against all that is taught about hygiene. Bacteria are commonly associated with three "D's" of dirt, disease, and death. True, this applies to some bacteria, but they are only a minority. Nevertheless, to add a little drama, let us consider these three D's starting with the first: dirt. The most dangerous bacteria hiding in dirt are possibly those causing tetanus. *Clostridium tetani*, as they are called, normally live in soil. A deep skin wound that is contaminated with soil or dirt poses a serious risk for tetanus. The disease is also known as lockjaw, as the unfortunate patient suffers from uncontrolled spasms, often starting with the jaw muscles but progressing into spasms of the back, causing a typical arched posture. The culprits are not the bacteria themselves, but a toxin they produce, which is one of the most potent biological toxins known to mankind. A vaccine was developed as early as the 1920s, and nowadays, most people in developed countries are vaccinated against tetanus. The shot has to be refreshed every ten years to provide complete protection. Since effective vaccination has been in place, the numbers of tetanus cases have declined, but it still occurs in developing countries at significant numbers, and even in developed countries there are regular cases, notably in nonvaccinated individuals. *C. tetani* is present in many soils, meaning that vaccination programs remain essential to keep the disease in check.

The discovery that many infectious diseases, to continue with the second D, are caused by bacteria was made in the nineteenth century. Great names in microbiology have helped establish the causal relationship, such as Louis Pasteur (1822–1895) from France, who provided convincing evidence that microorganisms did not appear spontaneously (as was commonly believed in those days) but had to be introduced to a media before they could grow. His name lives on in the term "pasteurization," which is the heating under pressure of a liquid to 120°C to kill off any bacteria present. Pasteur developed, as part of his bacterial research, a crude vaccine against the bacterial disease anthrax. Robert Koch (1843–1910) from Germany also worked on anthrax, as well as on tuberculosis, another serious bacterial disease. His most important achievements are his postulates, which describe a set of criteria that have to be met for a microorganism to be considered the cause of an infectious disease. These postulates are still applied in modern medical microbiology. Later, Alexander Fleming (1881–1955) from Scotland discovered penicillin, and Selman Waksman (1888–1973), a Ukraine immigrant to the United States, discovered actinomycin and streptomycin. These antibiotics are all still in use to treat bacterial diseases.

The third D stands for death. The most deadly infectious diseases, in terms of numbers of victims worldwide, are respiratory infections, diarrhea, HIV/AIDS, and tuberculosis. These four diseases take a toll of nearly ten million deaths per year. Of these four, tuberculosis is exclusively caused by bacteria. HIV/AIDS is a viral infection, but patients frequently die of secondary bacterial infections that their weakened immune system can no longer fight off. The other two main man-killing infectious diseases can be either of bacterial or viral cause. In the case of lethal respiratory infections, bacteria probably win over viruses. Of the over four million lethal respiratory infections globally per year, three million die from bacterial pneumonia, whereas seasonal influenza (a viral infection) is responsible for "only" half a million deaths per year. Diarrhea mostly kills children, and here viruses may actually take the lead, although cholera can claim many victims at once in large outbreaks.

Infectious diseases can also be ranked in terms of how likely one is to die once infected. This is not as straightforward as it sounds, as the clinical outcome of an infection may vary between individuals, depending on the dose of bacteria or viruses to which the person was exposed and on his or her immunological status and general state of health. When a person's immune system is not working properly, bacterial and viral infections become both more frequent and more severe. Immunocompromised individuals can suffer from this condition because of genetic defects, certain diseases, or as a result of required medication, for instance, after they have received an organ transplant. Infants and young children, especially when undernourished, and the elderly also have a higher chance of dying from an infection. These individuals may even suffer from infections caused by bacteria that do little harm to the rest of the population. There are a number of bacteria that cause disease only if the immune system of their host is weak. Despite this variation, there are a few infections that most people are unlikely to survive, once infected. Here, viral infections clearly win, with rabies being the most macabre example. Without vaccination, hardly anybody can recover from rabies, unless they rapidly receive antibody treatment. Once the virus starts replicating inside you, there is no hope and no treatment.

With these many dangers lurking from the bacterial (and viral) community, we could easily forget that the vast majority of bacteria have nothing to do with us. Most bacteria just live their short or long lives and completely ignore us. Some treat a human being as a warm support to grow on or in, in which case we call the individual a "host." In the majority of cases, the host does not suffer from the fact that he or she is used as a home. Bacteria living this way are collectively known as *commensals*. A few bacterial species cause harm when present, with a diseased host as the outcome, and now we call this bacterium a *pathogen*. In many cases, disease could be considered collateral damage, and other hosts, who were luckier, may have harbored the same bacteria without getting sick. Relatively few bacteria are completely dependent on the harm they cause for their survival. These pathogens will always cause disease to the host they infect, as the symptoms they cause are essential for their growth. Nevertheless, from the perspective of most bacteria, causing disease is not of main importance in the range of possibilities they have to survive and multiply.

Bacteria, like all other living organisms, live to multiply. They will produce offspring as long as conditions allow, and they will adapt their lifestyle to the local conditions that apply, as long as this is within their capabilities. Some bacteria have a very limited repertoire of lifestyle possibilities, so that you always find them living in more or less the same conditions, whereas others are real universalists and can be detected in a variety of environments. It would be silly to treat bacteria in general terms only, pretending they are all alike. A zebra is not very "typical" of all animals, especially if it has to serve as an example for insects, worms, and squid, as well as mammals. Likewise, *E. coli*, which is probably the most generally known bacterial species, is not "typical" of all bacteria. We can only pay respect to the true nature of bacteria if we recognize their diversity.

There is one thing we can learn from all this. If life exists on other planets of the universe, and if that life is comparable to what we know, we are most likely to observe it with a microscope. Bacteria lived on earth before we did and will do so after we no longer exist. The chance that some other planet is covered with bacteria is higher than the chance of finding intelligent life in outer space.

2

Tree of Life: Let Three Live

Before the stage is reserved for bacteria, a few concepts need to be explained, which are easier to understand with examples from the macrobiological world. All living organisms visible without the need for a microscope can be divided into plants, animals, or fungi, which are recognized as three biological *kingdoms*. These kingdoms can be further split into major groups. For instance, within the kingdom of animals, we recognize the five groups of vertebrates: mammals, birds, reptiles, amphibians, and fish. Those are all animals with a spinal chord. Each of these vertebrate groups again consists of subgroups, to be split up again, until one reaches the level of species or even subspecies. This scheme of categorization can be graphically represented as a tree with a stem from which major branches separate that split into minor branches, twigs, and so on. Already Charles Darwin (1809–1882) recognized the treelike organization of life, and the only drawing in his famous book *On the Origin of Species* is in fact a sketch of a tree.

Obviously, vertebrates are not the only existing animals. The vertebrates form a side branch, or *phylum,* within the kingdom of animals, which is called *Chordates*. If we follow this branch of Chordates along the tree toward the trunk, the branch merges with that of other animal phyla (other bigger branches), such as the Echinoderms (a phylum containing the spiny-skinned animals to which sea urchins and starfish belong), the Hemichordates (a phylum containing particular types of worms), or the Arthropods. For the sake of this example, we stop here and turn around, now moving in the other direction, toward the twigs of

Bacteria: The Benign, the Bad, and the Beautiful, First Edition. Trudy M. Wassenaar.

the phylum of Arthropods. This phylum branches out into insects with six legs (Hexapods), crustaceans (crabs, lobsters, shrimp, and their like), millipedes and centipedes, eight-legged arachnids (spiders, ticks, scorpions, and mites), and a few others. At this level of grouping, the group of six-legged insects contains the highest number of representatives of all land animals, whereas crustaceans rule the seas.

The tree of life represents a lot of information. Two species that are positioned closely together on the tree are likely to share more genetic characteristics than two species further apart. The tree not only illustrates genetic but also evolutionary relatedness. If we were to follow the tree from one species to a closely related one, we might have to pass only one branching point, called a *node*. This node represents a moment in the evolutionary past when their most recent common ancestor split up into the two relatives that now form two different species. To go from a ladybird to a tarantula, however, requires moving along the insect branch deeper into the tree, passing many nodes, until the node representing the most common ancestor of all spiders and insects is reached, and the arachnid branch can be followed toward the tarantula at the tip of one of its many branches. Each branching point thus represents an evolutionary time point at which a species split up into two, each of which then went their own evolutionary path. Fossils can also be placed on the tree of life, and "missing links" can be predicted even if their fossilized bodies have not (yet) been discovered.

Each of the major branches that "bear" many species has its own name, and these are always derived from Greek or Latin words. In our example, "phylum" (plural phyla) means race and "chord" in the word Chordate means gut or string, to emphasize the presence of a neural tube. Echinoderm is derived from the Greek words for sea urchin and skin, and Arachne is Greek for spider. Latin, with many (Latinized) Greek words, for a long time, was the lingua franca of science. It has left us with many biological terms that may be difficult to spell or even remember, but they are also melodious witnesses of historical biological developments and insights.

All biological species are also named in Latin, which avoids the Babylonian confusion that would arise if common names in multiple languages would be used. Latin biological names always consist of two parts. The first, capitalized name is the *genus* (plural genera, literally meaning race), and the second name gives, the *species* (both singular and plural). Latin names are printed in italics by convention. This way of naming, with a "general" description (genus) followed by a "specific" one (species), was already practiced by Aristotle. It compares to the given name and family name of individuals in some Asian societies, where the family name of individuals is given first, followed by the given name. In Western societies, the order is reversed, with the given name placed before the family name.

It was the Swedish botanist Carl von Linné (1707–1778), better known as Linnaeus, the Latinized form of his name, who standardized biological nomenclature. He ranked, to the best of his knowledge, plants into species and genera, and those again into the higher levels of families, orders, and classes. He proposed, in his

Systema Naturae, three kingdoms, that of plants, animals, and minerals. Although the last obviously does not belong to the living world, his proposal for naming and classification of plants and animals stuck. What Linnaeus started as a short pamphlet (the first edition had only seven pages) developed into a growing and never-ending project. Although many of the details have been modified since his first attempts, his system is still in use today. His fascination with ordering things is reflected by the word "taxonomy" (from taxis, order), which is what this field of science is called.

The genetic relatedness of organisms closely positioned on the tree of life does not always predict a similarity in lifestyle or habitat. Cats are more related to lions, and dogs to wolves, than cats are related to dogs, even though cats and dogs may share the same household. Body shape, or morphology (morphe means form), is a more reliable marker for relatedness than lifestyle—although even similarity in build can be misleading. A swift and a swallow may look similar, but they are not related, and a lizard may look like a miniature crocodile, but they are only very distant cousins. We can tell by looking at other characteristics, such as skeleton build or genetic characteristics.

The twigs on the outer branches of the tree of life, representing the relationships of present-day animals (and plants), are for the most part well established, but as we move further inward, the positions of the nodes become less clear. Very deep inner nodes represent ancient common ancestors, of which there may at best be incomplete fossil records left, or no traces at all. In this case, genetic evidence is used to establish the relative relationships of living species, but the interpretation of such genetic data leaves room for disagreement. Multiple and opposing interpretations may exist of what would be the "best fit" for a particular section of the tree. This is the reason why there is no scientific picture of the tree of life in this book, as it would only be "a" tree of life, representing one interpretation where several others might fit just as well. Even the branch ending in our own species, *Homo sapiens* (meaning "wise man"), contains twigs that are subject to hot debates. Fossil discoveries of our early ancestors frequently lead to scientific disputes about the exact position of nodes of our branch on the tree of life. These disputes concentrate on details only, whereas the large picture, the general grouping of hominids within the Primates, is beyond doubt.

All eukaryotes together, containing the kingdoms of animals, plants, and fungi, form a separate section, or *domain,* on the tree of life, called the *Eukarya*. However, the domain of Eukarya contains more than the macroscopic plants, animals, and fungi: there are many eukaryote organisms that live as single cells, and these unicellular eukaryotes are also included in this domain. All these diverse organisms share one property: their cells contain a nucleus, the compartment within their cells where the DNA is stored, surrounded by a membrane, like a little cell within a cell. This is what sets them apart from bacteria, whose cells do not have a nucleus. Bacteria are described as *prokaryotes,* a word that literally means "before a kernel," to emphasize the lack of a nucleus. Prokaryotes are always unicellular and microscopic and lack a nucleus. Eukaryotes always possess a nucleus; they can be unicellular or multicellular and exist as microscopic or macroscopic organisms.

The prokaryotic world forms another domain, or, in fact, two, as we will see. The domain of Bacteria (in respect of the highest level recognizable in the tree of life, a domain is written with a capital here) can also be represented as a tree with a number of major phyla that each branch off into further divisions. Few taxonomists will dispute the recognition of the major bacterial phyla (although some do not accept the word "phylum" and prefer to call these major branches "classes" instead); which species and their genera belong to which phylum is also clear, although mistakes have been made in the past and occasionally corrections are put in place. However, as we move deeper toward the base of the bacterial domain of the tree of life, uncertainty increases. The relationship between the various bacterial phyla is not always clear. There are a number of reasons for this disagreement, of which the absence of fossils is an obvious one. A further complication in bacterial taxonomy is that the definition of a bacterial species is not the same as that of a plant or animal species, as bacteria do not reproduce sexually. The degree of diversity within a bacterial species is also not constant. Some bacterial species contain isolates that are always genetically very similar, whereas other species display extensive genetic diversity, so that the ordering that Linneus had in mind is somewhat chaotic. Disagreement is far more frequent for the exact placement of particular bacteria into species and genera than it is for animals and plants. Quite a few outer branches have been misnamed or misplaced, in the opinion of (some) current scientists, and to this day, bacteria are being reshifted along the tree.

Occasionally, a "newly discovered" bacterial species happens to have already been described by an obscure scientist in the past, who gave it a name since long forgotten. In this case, after extensive discussion and review, it can be decided to use the oldest name available, which means that a species can suddenly change names. More frequently, a new name results from novel scientific insights revealing that a particular bacterial species should belong to genus B rather than to genus A, which changes its first name. Alternatively, it can be proposed that two "species" are really one and the same, or that an existing species better would be split up in two, which would change their second name. Such insights thus result in a new genus name or a new species name or even both, for a bacterium that has not changed its properties. It can make the life of a bacteriologist quite difficult, as if your acquaintances have suddenly changed names and their entries are now to be found on different pages of the telephone book. Changed bacterial names complicate the checking of scientific literature. Understandably, such name changes can receive fierce opposition and sometimes things are left as they were, for the sake of peace.

The apparent "mess" of the bacterial domain is understandable. From the examples of the macrobiological world, it is clear that lifestyle or morphology is only of limited use to establish relatedness, and many bacteria look more or less the same under a microscope. So how should we group bacteria, if not by their looks and behavior? In the old days, when research was dedicated to medical microbiology, distinctions were frequently made based on the diseases bacteria could cause. This has led to some inaccurate classifications that we live with even today. For example, shigellosis is a type of severe diarrhea caused by *Shigella* species, for instance *Shigella dysenteriae*, which, by objective criteria, are just particular nasty brands of *E. coli* (the "*E*." of *E. coli* stands for the genus *Escherichia*). There is no scientific reason to grant *Shigella* bacteria their own genus name, but taxonomists have not renamed *Shigella* bacteria to be incorporated into the *Escherichia* genus—yet.

In addition to disease-related descriptions, particular bacterial characteristics can be tested in the laboratory to establish similarities that are indicative of relationships. A wide variety of tests have been developed, which are frequently based on the type of nutrients a bacterium can grow on, the kind of waste products they produce, or how much variation in growth temperature they tolerate. Bacteria displaying similarity in these properties as well as in their morphology, as far as could be judged with the help of a microscope, were considered closely related. Nowadays, the major basis for bacterial taxonomy is the genetic material of the organism. The DNA is isolated from the cells and chemically compared. In addition to the chemical composition (which varies considerably between bacteria, as explained in Chapter 16), the DNA sequence of particular genes is also used as a criterion for grouping. One gene in particular, which is found in all bacteria known to date, is used as a marker for similarity and has served as an important indicator of taxonomic relationship. But even that sometimes causes conflicts. Now that we can read the DNA sequences of complete bacterial genomes, and not just that of one or a few genes, it is realized that a number of our previous insights were not very accurate. The repositioning of branches within the bacterial domain is an ongoing process.

The tree of life with the two domains of Eukarya and Bacteria is still incomplete. The prokaryotic world consists of two very large and different groups, *Archaea* and *Bacteria*. (Sometimes the term "eubacteria" is used to distinguish the latter from the other major group that is also known as "archaebacteria".) Archaea look like bacteria, live like bacteria, and for a long time were thought to be bacteria (and some microbiologists still support this view). But their properties indicate that they should be grouped on a very deep branch, separate from the other, true bacteria: they form the separate domain of Archaea. This gives a tree with three domains: Archaea, Bacteria, and Eukarya.

That archaea are different from bacteria became apparent only 40 years ago. Their membranes are chemically different from those of eubacteria, and some telltale genes, involved in very basic processes of life also differ between archaea and eubacteria. Since the membrane (which is made up of lipids) separates the inside of a cell from its outside, even the most ancient living cell must have had a membrane. When archaea and eubacteria have membranes of different compositions, however, it is difficult to envisage a common ancestor. For this reason, and for some of their important genetic differences, which become apparent only to the eyes of the specialist (for instance, the two groups use slightly different ways to reproduce their DNA), these prokaryotes have been given their own domain.

The name archaea implies that they are very old (the Greek "arkhaios" means ancient), but we do not know whether they existed before bacteria, arrived later, or developed simultaneously. When their name was proposed, it was believed that archaea were the first inhabitants on earth, and this idea is still propagated by some. Alternatively, it has been proposed that bacteria came first and archaea formed from these, whereby their different membrane composition had to be explained. In another view, an ancestral form of life that was neither archaeal nor bacterial, but something in between, split up into bacteria and archaea. These uncertainties are another reason why the tree of life, with its three domains, cannot be represented accurately on these pages. Some scientists even propose that life originated twice, independently, once producing the ancestors of what we now call archaea and once those of eubacteria. Maybe our tree of life indeed had two roots, but as seen in Chapter 6, archaea and eubacteria both contributed to the evolution of eukaryotes. It is reassuring that there are not two or three separate trees but an intermingled growth of branches that frequently split but sometimes merged. It means that somehow our distant past still records our relatedness to bacteria and archaea alike.

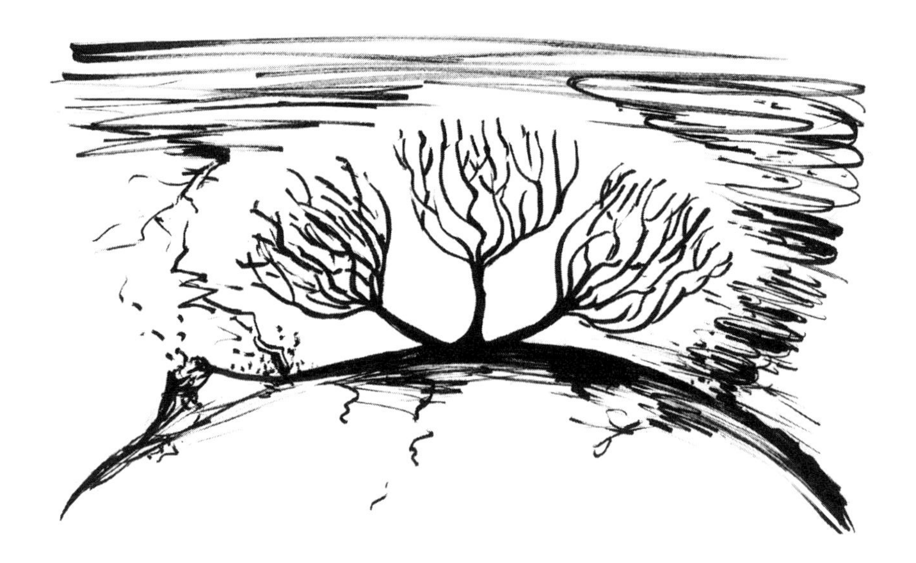

Within the domain of Bacteria (and now eubacteria are meant), a distinction can be made based on a laboratory staining test that has been in use since 1884.

It was developed by Hans Christian Gram (1853–1938), a Danish bacteriologist, while he was staying in the laboratory of the famous Paul Ehrlich (1854–1915) in Berlin. The test combines the use of two dyes, one violet and the other purple. Bacteria are fixed on a glass plate by briefly running it through a flame. First the violet solution is added, then the purple one, each step followed by a few washes. This procedure colors the bacteria either bright red-violet or dark purple, which can be seen when the glass slide is mounted on a microscope. The red-violet bacteria are called *Gram negative*, and the dark purple ones are called *Gram positive*. The test is simple, fast, and inexpensive and happens to depend on a characteristic that we have already seen is very important: the difference is due to their membrane. Although all eubacteria have membranes of similar lipid composition, Gram-negative bacteria are surrounded by two membranes, whereas Gram-positive bacteria have only one (and thick structure mainly consisting of a polymer called peptidoglycan). Exactly how Gram negatives got their two membranes is not known, but the distinction is important and consistent. Only a few bacteria do not fall into these two major classes, being neither Gram positive nor Gram negative. For most bacterial examples mentioned in this book, their Gram type is given.

The membrane is the only "structure" that shapes the bacterial cell. It functions as skin and skeleton and mouth and nerves, all in one. A Gram stain not only separates bacteria into different taxonomically deep divisions but also can predict, to some extent, which antibiotics will kill them and which would not, as antibiotics frequently interfere with membrane building. Chapter 12 elaborates on this some more. Hans Christian Gram is not among the "champion league" of bacteriologists, but his name appears in nearly every scientific publication on bacteria. To write it without a capital (and hence cause confusion with the standard unit of mass) is not doing justice to the value of the test he developed.

The tree of life, with the three domains, multiple kingdoms, and even more phyla so far mentioned, is still very incomplete. The fungi have hardly been described, to which molds, mushrooms, and yeasts belong. Mushrooms are clearly multicellular organisms, whereas yeasts are unicellular and moulds can be either. They are all eukaryotes because their cells have a nucleus, and they form another kingdom within the Eukarya domain. In contrast to what one would think, most eukaryotic life is actually microscopic, despite the abundance of plants and animals. These single-cell eukaryotes are sometimes indignantly bundled together as "protists," and this unofficial term includes the dinoflagellates, amoebae, diatoms, protozoans, and many others, which all have their own branches. The kingdom of plants should not be forgotten, to which algae belong. Even the animal kingdom is still incomplete without mention of the nematodes, sponges, snails, jellyfish, different classes of worms, clams, and many other animal life forms. Protists, plants, and animals hardly play a role in this book, unless they serve as a host for bacteria. The bacterial branches, for a change, receive more attention than their macrobiological counterparts that fill the pages of most books.

Taxonomy is the scientific discipline to order organisms into groups, so that members within a group are more related to each other than to organisms of another group.

- Taxonomic groups exist at various levels. As the level increases, more members are included and their genetic distance increases. Groups at a lower taxonomic level contain fewer members that are more closely related.
- The highest possible level of taxonomy is that of a **domain**. There are three domains: Eukarya, Bacteria, and Archaea.
- **Eukarya** is the domain containing all **eukaryotes**. These are organisms whose cells have a nucleus, a membrane-enclosed compartment within the cell containing most of their DNA. All multicellular organisms (plants, animals, and fungi) are eukaryotes, but so are unicellular fungi and protists. These eukaryotes live as single cells, and thus are microbes, as one can observe them only with a microscope, but they are not bacteria.
- The domains of **Bacteria** (also called Eubacteria) and **Archaea** (sometimes called Archaeabacteria) both contain exclusively single-cellular organisms that are all **prokaryotes**. Their cells do not have a nucleus; instead, their DNA is localized in a condensed area within the cell.

In **bacterial taxonomy**, the highest level of grouping is a phylum and the lowest is that of a species.

- A phylum consists of multiple "classes." At the next level down, several "orders" can be found within a class, and within orders we find the families. Every family consists of multiple genera, and finally, these are divided into species.
- As with plants and animals, the lowest official taxonomic division of bacteria is a **species**, although subspecies are recognized for some bacteria.
- The **Latin name** of all living organisms, including bacteria, is given by its **genus** name followed by its **species** name, the lowest two taxonomic divisions recognized. From their names, one cannot tell to which phylum bacteria belong.
- The Archaeal domain is the least characterized. Archaea can also be grouped into deep branches, although the term "phylum" is less frequently used for archaea. Some archaea have not been named at all, being given a number only, but they are so different to all other known archaea that they likely represent novel phyla for which we only have single example organisms.

3

How Old Are They?

It is hard to say how old an individual bacterial cell might be, as there are no obvious signs of aging. In contrast to human cells, bacteria can produce offspring that will keep dividing indefinitely, as long as the conditions allow growth (normal cells of a human individual will eventually stop dividing, although cancer cells keep growing). Most bacteria multiply by binary fission, which divides a growing cell into two smaller daughter cells, although some species use budding or can even produce multiple offspring simultaneously. With binary fission, half of the original cell is maintained in each daughter. It means that each and every cell contains some material that was already present in its mother cell, as well as material that is newly made. This complicates the definition of "age," but we can at least consider how long a cell can survive without dividing or dying.

The maximum age of an individual cell depends on the species. Bacteria have evolved various methods to survive harsh conditions. In general, their first rule is: do not waste energy. Cells that encounter suboptimal conditions usually switch off any process that requires a lot of energy and is not essential, given the circumstances. This means that bacteria stop dividing when it gets too hot or too cold for them, when the amount of nutrients becomes limited, when toxic compounds build up, or there is too little humidity. Bacteria "sense" these factors and, in response to these alarm signals, stop growing. This argument can be turned around by saying that cells divide only when the circumstances allow them to do so. Otherwise, the cell will attempt to repair damage to its major constituents, which are proteins, lipids, complex sugars, and nucleic acids (to which DNA belongs). It will retain

Bacteria: The Benign, the Bad, and the Beautiful, First Edition. Trudy M. Wassenaar.

its salt balance and keep the energy carriers at a desired level, but it will reduce all energy-consuming processes to a bare minimum.

Scientists have kept cells of *Vibrio* (a genus of Gram-negative bacteria to which the dreaded cholera bacteria belongs) in sea water for one year, during which the cells did not multiply but maintained their essential processes: they were still alive. In the presence of sediment, the cells survived for six years. Some species can easily beat this: they can switch to a vegetative stage in which they literally shrink to only a fraction of their usual size. All processes are switched off, and one could say they are dead. This state is called an endospore, or *spore* for short, a somewhat misleading term, as it has nothing to do with reproduction as is the case with fern or fungal spores. Quite the opposite, bacterial spores are masters in survival but do not easily multiply. Only when the spores encounter favorable conditions that allow growth again, they will revitalize themselves, boot up cellular processes, and return to the land of the living.

Bacterial spore formation is a survival strategy for sitting it out, waiting for better times to come. This can take a long time—very long indeed. *Bacillus* species and some other Gram-positive bacteria belonging to the phylum Firmicutes are notoriously good at forming spores that survive for extremely long periods. This has been known for decades, and at one time, it was thought spores could be used in battle. In the mid-twentieth century, several countries experimented with *Bacillus* species (among other disease-causing bacteria) to be used as a military weapon. Biological warfare was seriously considered as a useful addition to existing ammunition, and *Bacillus anthracis*, the cause of anthrax, was one of the investigated candidates. Its ability to produce spores that are very persistent, easy to generate, and very infectious on inhalation was deemed a positive trait.

During the Second World War, in the United Kingdom (among other countries), military scientists experimenting with anthrax bacteria decided to do a test: they confiscated Gruinard Island off the coast of Scotland and brought in a number of sheep. The island was then deliberately contaminated using anthrax-loaded bombs. Sure enough, the animals got sick, and the first ones died within a week. The experiment had been a "success." Less successful were attempts to clean up by setting fire to the vegetation. Decontaminating the island failed miserably, and the whole site had to be kept in quarantine. Twenty years later, the island was still unsafe, as anthrax spores that could revitalize themselves when eaten by sheep were still present, with the deadly disease as the undesired outcome. A new attempt at decontamination was carried out in 1986, spraying large amounts of formaldehyde (a very potent disinfectant) dissolved in seawater over the island and removing a layer of topsoil from the most heavily contaminated locations. This helped, and the island was declared safe in 1990, 48 years after the deliberate spread of spores. Without that intervention, the spores would probably have survived for centuries.

For those bacteria that are unable to form spores, low temperatures enable longer survival than usual. Scientists keep stocks of bacteria in freezers at −20°C (−4°F) or submerged in liquid nitrogen, which is below its boiling temperature of −196°C (−321°F). Usually, the bacteria are protected from frostbite by adding glycerol to the liquid in which they are stored, and under these conditions, the cells survive without any measurable loss, as long as freezers do not break down, power outages do not occur, and the liquid nitrogen is regularly filled up.

Freezing temperatures also conserve bacteria in the natural world. Bacteria have been isolated from permafrost soil that must have been frozen for tens of thousands of years, after which they could still be brought back to life. Glacial and polar ice have also been hiding places for very ancient bacteria. It has been demonstrated that some bacteria have survived no less than a million years in polar ice—which would easily make them the oldest individuals on earth. Sediment grains that are coated with a very thin layer of salty water also provide a tiny habitat where bacteria can survive for thousands of years and beyond, and they will even have nutritious minerals to feed on. Bacteria trapped in amber are less lucky, but even these have survived for hundreds of thousands of years. However, such locked-up bacteria are not immortal. Eventually, they will die from lack of nutrients or accumulative irreparable damage of their DNA due to natural radiation.

The question "how old is that man?" produces an answer of a different magnitude when compared to "how old is man?" The life span of an individual is negligible compared to the life span of a species. When individual bacteria can live a million years, how old could their species be? An intriguing question that we cannot answer with certainty in the case of the ice-recovered species, but we do know that the age of different bacterial species obviously varies. Bacteria that can live only in man (of which there are quite a few) must have evolved after man did, so they cannot be older than our species. Bacteria causing sexually transmitted diseases are completely host specific, with different species living in different hosts, and they can multiply only inside that host. So what happened to the bacteria specific to our ancestors, or those dependent on the dinosaurs? They may have become extinct with their host, or, if they were lucky, some of their offspring may have adapted to live in an alternative host and subsequently evolved into new species.

Life forms do not remain constant over time; they slowly and gradually evolve and change, until a point is reached where the existing form is clearly different from its ancestor: a new species can be recognized. With plants and animals, the definition of a species is relatively strict. As a general rule, when two individuals can breed and produce fertile offspring, they are grouped within one species. This rule cannot be applied to bacteria since their cells reproduce asexually. As explained in the previous chapter, bacterial species are recognized by differences in their characteristics, but within a species, some variation may exist. The degree of variation between isolates of one species is again variable: some species allow more variation between individual isolates than others. It is not always clear how much variation is required or sufficient for a new species to be defined. Some bacterial species display a lot of variation between isolates, whereas other species seem to consist of completely identical individuals. The biological distinction of species is a bit artificial when dealing with prokaryotes, but without this concept, the microbial world would be even harder to describe.

In addition to the difficulty we meet when defining bacterial species, their evolution is faster than that of animals and plants because they have had more time to evolve (existing so much longer). In addition, their generation times are much shorter. The fastest growing animals may produce offspring within days, whereas many bacteria can do so within hours, and some can multiply in less than ten minutes. A few mutations are likely to occur every once in a while within a population, which by themselves are evolutionarily insignificant, but eventually their accumulation drives the changes necessary to adapt to novel environments, resulting in the development of novel species. The build-up of mutations over generations occurs much faster in bacteria than it does in animals and plants. Even animals and plants living today differ from those that lived in a distant past, so early bacteria must have been quite different from those we know today. Nevertheless, this is often ignored when the bacteria that lived in past eons are considered; lacking specific knowledge about these past inhabitants, we describe them in terminology only fit for present-day bacteria.

We do not know what bacteria looked like when the earth was still young. Given the conditions that applied then, they must have been able to live without atmospheric oxygen and to endure extreme temperatures. They may have lived in water before exploring land, but all of this is uncertain. For many present-time archaea, oxygen is toxic, and some can live at temperatures much higher or lower than room temperature. As discussed in Chapter 15, some can endure high concentrations of salt or other minerals or, instead, other conditions that may have applied thousands of millions years ago. Maybe these present-day inhabitants of extreme conditions are distant cousins of the archaeal life forms that seeded our planet with life, but we do not have a way to go back and check.

The early earth provided the conditions in which life could develop, but once life existed, it changed the conditions on earth. When the earth was young, its atmosphere was mainly created by volcanic activity and as a result had a different composition from that of today. Methane, ammonia, and sulfur dioxide were probably abundant in the air, while oxygen was absent. We know this from geological deposits that were formed during this period, containing minerals such as pyrite (fool's gold), which consists of forms of iron and sulfur that could not have been stable in the presence of oxygen gas.

It is just as well that the earth's air was free of oxygen, as the building blocks of living cells could not have been formed in its presence. Amino acids, with which all proteins are built, can form spontaneously from chemicals naturally present on the early earth through simple chemical reactions, but only in the absence of oxygen. Although proteins are essential to all living cells, without the presence of the nucleic acids DNA and RNA, a cell could not exist. RNA seems to have played a more important role for early life than its sister molecule DNA, which may have entered the stage later. Nobody knows for sure, but a suitable explanation of how this RNA came into existence has been proposed. How exactly life was shaped from abiotic chemical constituents is still a bit of a mystery, and whether it only happened once or suffered a few failed starts before kicking in is unknown. One thing is certain: once life was abundant, it changed the atmosphere gradually to enable further life forms to evolve.

Whether from deep-sea sedimental rocks or continental deposits, all geological evidence suggests that oxygen started to accumulate in the atmosphere during the Proterozoic, a long period starting 2500 million years ago and lasting nearly 2000 million years. This oxygen must have been produced by bacterial activity through photosynthesis: living cells converted carbon dioxide into biological material, with the formation of oxygen gas as a by-product (photosynthesis is explained in detail in Chapter 13). The vast increase in atmospheric oxygen was the result of trillions and trillions of bacteria doing their photosynthesis job, but whether this was a gradual or an irregular increase is disputed. Eventually, the atmosphere contained so much oxygen that it allowed the formation of life forms that thrived on oxygen, instead of being poisoned by it.

In most textbooks, Cyanobacteria (which form a phylum of their own) are given the credit for producing the oxygen that life is now dependent on. This is understandable, as even in present times, Cyanobacteria are responsible for a vast amount of oxygen produced every day. These Gram-negative bacteria were formerly known as blue-green algae, but they are true bacteria, whereas algae are eukaryotes.

(Some biologists still consider Cyanobacteria as part of the plant kingdom, but most microbiologists agree that they are bacteria). In the first chapter, it was already mentioned that Cyanobacteria can form stromatolites. These are thick mats of bacterial cells deposited in shallow coastal waters that appear strikingly similar to the fossilized stromatolites that have been found to be 3600 million years old. However, it is unlikely that those ancient bacteria creating stromatolites, and producing oxygen as they grew, were the same as those living in modern-time stromatolites, without having undergone any evolutionary changes. The microorganisms responsible for the oxygen in our air may well have been the ancestors of Cyanobacteria, but almost certainly they will have been different from today's living oxygen producers. The basic chemical processes they carried out may have stayed the same, but their details will have differed from what we see today.

Geologists are used to deal with events that took place millions or more years ago, but it is hard to imagine just how long these periods are. We can envisage the landmarks in biological history better if we think of a time line, on which we project time as if it were a distance. Since the cells of bacteria measure rarely longer than a hundredth of a millimeter, we will make that length represent one year. A millennium would be given by a thousand of such bacteria aligned head-to-tail, which would span a distance equivalent to 10 mm, and a million bacteria, representing a million years, would cover 10 m. On this distance scale, the beginning of Earth is 47 km away. The oldest bacteria evidenced by fossil stromatolites can be found at a distance of 36 km, and around 25 km, the oxygen started to rise in the atmosphere. The oldest single-cell eukaryotes are possibly 21 km away. When the first multicellular organisms arose is disputed and could be soon after the appearance of the first eukaryotes, but there is good evidence that simple animals and plants existed at a time represented by a distance of 6 km. The dinosaurs lived in a period that is only a small distance on our scale, and they became extinct at 65 million years ago, which is only 650 m away. At that time, mammals already existed, which appeared 200 million years ago, equivalent to 2 km. The genus *Homo* appeared very recently, at 25 m, and the first human-looking creatures lived 2 m from our toes. Compare these 2 m with the 36 km that represent the time bacteria existed, and one can see how much longer these little cells have existed compared to our likes.

The bacteria that populated our planet have endured quite some challenges. During its existence, the earth's climate has varied considerably, and life forms have come and gone. Even the atmosphere has not remained stable, with variations in oxygen and carbon dioxide levels that have severely influenced the biosphere and climate. The early earth froze a number of times completely, which must have been hard for any existing life, but as we have seen, bacteria can survive freezing conditions. Massive extinctions ended thriving ecosystems, making room for new and different life forms to evolve. Bacteria were always an essential component of these ecosystems, whether on land or in the seas. They evolved, just like animals and plants did, but, unlike their bigger co-inhabitants, they came and went without leaving a trace.

4

On the Move

The invention of the wheel was not a unique event—it was independently invented in multiple cultures. Rolling heavy objects over pebbles or logs could have preceded the use of true wheels, which are attached to an axis. Rotation is a very efficient means of converting energy into motion. So it must come as no surprise that biology also uses rotation to produce net movement of a body, although not in the form of a wheel but more like a propeller.

Many bacteria swim, using one or more rotating *flagella*. These are long filaments that look like a microscopic whip or tail. Bacteria can move forward by fastly spinning these tails. The flagella are attached to the bacterial body through a structure that is even called a motor. But how exactly do bacteria manage to build this complex structure, and how can they rotate it and move with it? The flagellar filament is made of protein. Proteins come in many forms, shapes, and properties. The proteins that make up the filament of a flagellum have such a shape that they spontaneously assemble into long chains, comparable to how magnetic rods would spontaneously form a chain, although the forces keeping the flagellar proteins together are hydrostatic and not magnetic forces. And instead of a linear chain, the flagellar proteins are stacked in such a way that they spiral up to form a thin and flexible tube, the flagellar filament.

The flagellar motor is also made up of proteins, different from those of the filament. The motor consists of an outer ring of proteins embedded in the membrane and an inner protein ring that rotates and to which the filament is attached. The rotation that is necessary to cause motion uses energy. This energy is produced

Bacteria: The Benign, the Bad, and the Beautiful, First Edition. Trudy M. Wassenaar.

by pumping ions (charged atoms) along the motor out of the cell. Since it takes energy to pump ions into a cell (where their concentration is higher than outside), such energy is released when ions are allowed to leave—and this energy is used to "push" the inner ring up to the next point where a new ion can pass. With every little puff of ions leaving the cell, the motor turns a bit, like the scoops on a watermill. The effect of this rotative momentum is increased by the long filament, and as a result, bacteria move forward (or backward, depending on what one calls the front end). The loss of ions has to be compensated by pumping ions back into the cell through the membrane using specialized ion pumps, at the cost of energy.

Many bacteria have flagella at both ends of their body so that they can move in either direction, depending on which flagellum is rotating. Bacteria with a single flagellum can switch the rotation from clockwise to counter-clockwise to change direction, or they just stop, wait till their body has drifted to a new position, and start swimming again. Others have multiple flagella sticking out at various points from their small body, which twist into one super-whip when they all rotate in the same direction. By suddenly changing the direction of rotation, the whip unwinds and their body tumbles, after which the individual flagella reassemble and the cell can swim in a new direction.

The movement of a single cell seems uncontrolled and random when viewed under the microscope, but bacteria do in fact keep a general direction. Motile bacteria (as those that move are called) will swim toward a food source and move away from repelling compounds. They sense these via signaling proteins in their membrane, a process called *chemotaxis*, further explained in Chapter 21. This causes them to swim in the desired direction, but even so, they do not proceed in a straight line. Typically, bacteria swim at top speed for a few seconds, then stop and tumble, possibly to reorientate themselves, and then continue in a slightly different direction. The overall speed can be about 1 mm/min, which seems to be slow by any measure. Consider, however, how small bacteria are. At top speed, they may move a distance covering 30 times their body length in a second. That would be of the same order as a man swimming at 50 m/s, or about 160 km/h (100 mile/h)! To be able to reach such speed, even for a few seconds, is truly remarkable.

Not all bacteria have flagella, and some species are completely nonmotile. However, looking at them under a microscope, they appear to be constantly moving, bouncing, shivering, and hopping. Their random movement is the result of a bombardment of water molecules. The movement was first described by Robert Brown (1773–1858), a botanist who loved the microscope. He discovered that plant cells have a nucleus, the dense kernel that, as was later found out, contains the cell's DNA. That discovery was a landmark, since, as already discussed, the absence of a nucleus happens to be the most important criterion for a cell to be called a eukaryote or a prokaryote. Brown also observed the random movement of cells on a microscopic scale, which is now known as Brownian motion.

Nonmotile bacteria have to get their food from their direct environment, while motile bacteria can swim toward nutrients. Motile bacteria are notably those that live in watery environments, and in the intestine of animals, whether for good (as the bacteria that digest our food for us) or for bad, causing infectious diseases. There is a good reason why intestinal bacteria are sometimes motile: the intestine is constantly moving, shoving its content along with the bacteria living in it. By swimming upstream, motile bacteria can prolong their stay inside the body. Similarly, our body uses a flow of mucus to remove bacteria from our airways, a flow of tears to clean our eye surface, and even the flow of urine to flush the urinary tract. Any bacteria that can resist such flow by motility have a better chance of multiplying at such sites, possibly with an infection as the unwanted consequence, from the host's point of view. However, not all intestinal bacteria are motile. Many bacteria withstand the constant flow of the body's fluid defenses by holding firmly to the surface: they attach themselves to the cells that form the mucous lining by means of short hairlike structures called *pili*. Attachment using pili is a strategy alternative to motility to gain a foothold at a site where food is plentiful. Many bacteria combine motility with adherence: they swim with their flagella to reach the desired surface and then hold on to this with their pili.

Pili are also made of proteins, of a type different from those that make up flagella. Pili are thinner and shorter than flagella, and they do not spin. Similar to flagella, they can be visualized with an electron microscope only, as both are too thin to be visible under a light microscope. Some bacteria carry so many pili on their cell surface that they look like little mops. There are various kinds of pili, one of which is called *Type IV pili*. These are probably very old, evolutionarily speaking, and they can be found on both Gram-positive and Gram-negative bacteria. Some of the proteins needed to produce pili even have a counterpart in archaea, where they function in producing archaean flagella, so it seems components of these two different structures have been mixed up in evolution. As their name implies, Type IV pili are only one of different types of pili that exist, and some bacteria can produce a variety of coiffures, whereby some pili are even curly. But none of these are found on such a diverse number of bacteria as Type IV pili, which can do a variety of things. They are not only used get a grip on surfaces but can also function as a sluice gate to secrete proteins out of the cell or let DNA enter the cell. And although they do not rotate as flagella do, having Type IV pili can even lead to motility.

Imagine bacteria having multiple "arms" with which they can grip tiny anchors one at a time, moving like a rock climber using hands and feet. Some bacteria use their Type IV pili as hands to move along. Again, bacteria living in our gut often have Type IV pili, even though they may have flagella as well. They swim while moving freely through the mucus, but having once grabbed the intestinal cell lining, they may use their pili to move along that surface.

Archaea can be motile too, and as was mentioned above, they have flagella just like true bacteria. That could mean that their most recent common ancestor already possessed flagella, in which case, these structures must have formed very early in evolution, since a common ancestor of both archaea and bacteria (in case this existed) must have lived when our planet was still very young. Instead of flagella already being present in their last common ancestor, it is also possible that one "stole" flagella from the other, as possession of flagella provides a huge advantage if you want to be motile. In this scenario, flagella may have evolved in eubacteria, and by the time all genes involved in producing flagella cooperated smoothly, they were taken up by archaea, who, in one swoop, would have learned how to produce flagella as well. As discussed in the next chapter, genes are sometimes exchanged between species, and even the exchange of a whole set of genes, all working to produce one structure, is not impossible. Nevertheless, it is unlikely that this happened, and there is an alternative and more likely explanation as to why flagella can be found both in archaea and eubacteria. Flagella may have evolved on independent occasions.

Scientists have discovered significant differences in the structures of archaeal and bacterial flagella, which indeed hint at independent evolution. It seems that the flagella of archaea may have evolved from prototype pili, different from the evolution of bacterial flagella. These different starts of their evolution resulted in slightly different end products, which nevertheless look surprisingly similar. This is an example of *parallel evolution*. Parallel evolution does not imply that two systems evolved at the same time but that independent processes produced similar outcomes. Parallel evolution is very common in both the microbial and the macrobial world.

It appears unlikely that evolution, which is the result of random mutation, produces the "same" product twice. Nevertheless, this can happen when that product is optimally fit for its purpose. The optimal way to swim, for a bacterium that needs to get around through water or mucus, balancing energy use with speed, is to have a rotating motor that spins a long filament, meaning that such a structure may outcompete other, less effective means. Flagella are simply superior to move bacteria around. Furthermore, there are constraints to the proteins that must build these structures, which may also contribute to a similar outcome of independent "inventions."

Some bacteria that cannot swim, but do not want to stay where they are, can glide from one location to another. Flagella enable swimming in liquid, whereas gliding bacteria move along a solid surface, but this gliding is independent of pili. Exactly how bacteria glide is not completely clear. Gliding is about half as fast as swimming: a glider could span half a millimeter in a minute. Some Cyanobacteria, for instance, *Synechocystis* species that colonize solid surfaces, use gliding to move around, and they leave a slime track as they go, like miniature slugs. It appears that they "spit" themselves forwards: by secreting slime under force, they move. This propulsion would require the production of slime, a temporary storage chamber, and a small hole through which the slime is forced. Since these gliding bacteria can steer themselves in a desired direction, the spitting must also be coordinated by sensors that tell them where they need to go. How this all works in the cell is still under investigation.

Propulsion is also the mechanism behind gliding Myxobacteria. These extremely interesting Gram-negative bacteria (of various genera but all belonging to the phylum Proteobacteria) can sometimes be found in animal dung or organic-rich soils. Myxobacteria are called *social* bacteria, as they form communities that work together. Despite that term, they are not very kind to other bacteria, as they are predatory by nature. They produce natural antibiotics to kill other bacteria, which they subsequently lyse (meaning they destroy their membranes) to feed on their cell content. As long as food is plentiful, Myxobacteria cells will traverse by gliding in all directions radiating out of the main population, the colony. But when food becomes scarce, they follow their own slime trail back to form a dense mass. What happens next is unique in the bacterial world. The long bacteria extend to form a small stalk upwards, and then the top cells start deforming to become a fruiting body, which can be brightly colored. More bacteria move up the stalk to the fruiting body, changing their cells as they do from long rods to round shapes. The round cells produce little heads on top of the stalk. Variation in the shape and function of cells depending on their location within a multicellular structure is called *differentiation* and normally occurs only in cells of multicellular eukaryotes. But Myxobacteria use a simple form of cell differentiation, and this may be an example of how the early ancestors of eukaryotes learned to produce multicellular organisms with differentiated cells to perform various tasks. The round cells of Myxobacteria inside the fruiting body, called *myxospores*, can survive lack of food, lack of water (desiccation is a serious condition for many bacteria), UV irradiation, and temperatures that would kill their normal cells.

Myxospores may not be as sturdy as the spores formed by *Bacillus*, but their act of moving together and collaborating to form a multicellular structure in which cells have different functions, depending on where they are located, is quite a feat for "simple" bacteria. The cells can even move in synchrony to produce "ripples," which are fascinating to watch, although we do not know why they do it. Somehow, it does not seem so strange that bacteria being able to both socialize and move have learned to "dance" together. Maybe they just do it to signal they are one happy family together.

5

Needles

The spray felt cold on my skin and was quickly swept away by the nurse, who, without hesitation, plunged a needle deep into my flesh. For an instant I worried she would hit a bone, but before I knew it, the needle was out again. A stinging sensation slowly developed into pain. A small Band-Aid completed the ritual, and while I put on my coat, I thought about the signals my body was now sending out to investigate and eradicate the foreign fluid that the injection needle had left in my arm's muscle. Soon, white blood cells would be recruited; blood vessels would relax and widen, causing a slight local swelling; and while nerve cells shouted at my brain that something was not right (which they did by sending pain signals, as they have only a limited vocabulary), other white blood cells were already mocking up the proteins that had deliberately been added to the vaccine, proteins that resembled those that could be present on the outside of a nasty influenza virus. Soon, antibodies that could recognize and bind to these proteins would be produced. Next time such a virus tried to gain a foothold at the inner lining of my nose or throat, so-called memory cells would recognize their protein coat and activate immune cells to produce more antibody and to destroy the virus particles before they could do much damage.

What an ingenious device, this injection needle. If the vaccine had been applied simply to my skin, it would have had no effect at all. Swallowing the liquid would not be an option either, as stomach acid and digestive enzymes would rapidly degrade and inactivate the useful proteins. Spraying the vaccine solution into the nose would also have had little effect, as the proteins would not reach the white

---

Bacteria: The Benign, the Bad, and the Beautiful, First Edition. Trudy M. Wassenaar.

blood cells that are essential to produce the desired immune response. But injection of the vaccine deep into a muscle, at a site where the targeted virus would never appear in real life, flared red flags to my immune system. Once the right immune responses are activated, it no longer matters where the virus enters: it will be attacked and dealt with.

The prick had hardly left a bloodstain on the Band-Aid that came off that night. For a few days, my arm felt stiff and the site of injection was sore, but that was a small discomfort to prevent worse: possibly a week or more in bed with flu, with a small chance of severe complications, a chance I would rather not take. The vaccine that was delivered by injection can only protect my body thanks to my immune system. The immune system is not a single organ in the sense that a heart or a liver is. Rather, immune cells are located everywhere where invaders such as bacteria or viruses most likely enter or do damage: a variety of cells patrol through the body via blood and live in lymph nodes and in the skin and the gut.

The immune system has two general modes of defense. An *innate response* is a defense strategy we are born with: when bacteria enter our blood stream, for example, specific blood proteins attack them by poking through their membranes to make the bacterial cells leaky and kill them. Viruses are mopped up by some white blood cells, specialized for this task, and are subsequently killed. As a result of immune cells producing particular signal proteins, the body temperature is raised, which makes it more difficult for bacteria to multiply. Thus, fever is one of the innate defense strategies of the body to fight infections. The innate immune system is always ready to react, but it does not learn from experience and has a limited repertoire. The *adaptive immune response*, on the other hand, is very specific and has a far broader repertoire. But it is slow, because it has to "learn" what enemies it should attack: cells have to encounter these a first time, and it will take some time before they have learned to take action. The production of antibodies, on which the action of vaccines is based, is part of the adaptive immune response. Antibodies are proteins that can bind with high specificity to a particular bacterial cell or virus particle, and for every type of bacteria or virus, novel and unique antibodies are produced. Their repertoire is basically unlimited, but producing antibodies has to be learned, which can take a few days. Once the immune cells have mastered it, they will "remember" what antibodies they need to produce. The next time the same bacterium or virus enters the body, specific antibodies will quickly be produced. The antibodies then bind to their target, so that other immune cells recognize these "marked" threats, and kill them.

Immune cells can kill, but they should not destroy body tissue. Therefore, immune cells learn to recognize our own body cells by recognition of signaling "flags" that all our cells carry, so that anything else lacking these flags can be recognized as "foreign." This way, killer cells leave our body cells in peace. When this regulation goes wrong, autoimmune diseases are the result: the immune system starts to attack our own cells, which can lead to an array of diseases, some relatively mild but others with devastating consequences. The signaling flags on body cells are unique to every individual, which is relevant in organ transplantation. Tissue implants from a different individual would quickly be recognized as carrying the wrong signals and would be destroyed by the immune system of the recipient. Only by careful matching of donor and recipient, based on the signals carried on the cells, and by suppressing the immune system with medication, can a transplanted organ be tolerated. An immune response also has to be tightly regulated, so that after a threat has been dealt with, it is tuned down. In the absence of any dangerous particles, the immune cells should remain silent. That regulation sometimes goes awry, and as a consequence, the immune system overreacts to substances that are completely harmless. This is what we call an allergic reaction.

Since immune cells are constantly patrolling our body, a vaccine, containing proteins specific to a particular disease-causing virus or bacterium, can be injected in a muscle and still be picked up. This is sufficient to set off the adaptive immune response, and a few cells will learn to produce the correct antibody, only to be ready for a strong defense when the real virus or bacterium would try to enter the body, independent of the site. Injection needles are scary to some people, but they provide an optimal means to trigger an immune response without the damage a real infection would cause.

The trick of a hollow needle to push proteins through an impenetrable barrier has been learned by some Gram-negative bacteria, too. The discovery of such bacterial "injection needles" that are known by the name *Type Three Secretion System*, or TTSS for short, has been a challenging puzzle that has not been completely solved yet. It is one of the most complex structures to be found in the bacterial world.

How did bacteria evolve this complicated device, the microbial equivalent of an injection needle? The structure is so complex, with so many proteins involved and all needed and essential to produce a functional TTSS, that one could be tempted to believe this is too complex to be the result of shear chance and evolution. The argument is not new. In mammalian evolution, the eye is used as an example of a complex structure that, according to evolution deniers, is too perfect to have arisen by chance. Critics of evolution deniers will argue that the eye is not perfect at all and any intelligent designer could come up with some major improvements. Moreover, it is not the product of chance only. The eye originated from preexisting structures; even some bacteria produce light-sensitive proteins. Such ancestral structures could evolve to optimize functionality, which was aided by combining other preexisting solutions to biological problems. At the same time, the development was hampered by any predecessor's structural shortcomings, and these shortcomings are still notable in the present-day eye. Mutations may arise from chance processes, but selection can build layer after layer of complexity and functionality. Likewise, the bacterial TTSS was not put together from scrap but was probably hijacked from an existing simpler structure, for which flagella are a possible candidate. Once it evolved, it could spread to other bacteria, but before dealing with that, we will consider what a TTSS is made of.

Essentially, a TTSS is a hollow tube through which material can be ejected. As with so many biological processes, proteins do the job. Many different proteins are involved to build a functional TTSS, and all these proteins, typically 30 or so, are coded by their own genes. The structural proteins that are essential to actually build the tube are not the only ones needed: additional proteins must combine the building blocks in the right order. There are proteins that build a bridge to cross the two membranes of the cell (only Gram-negative bacteria carry a TTSS, and these have two membranes), at the same time preventing cell leakage. Other proteins regulate when a TTSS must be produced and when it is not needed. Particular proteins have the task to fold the building block proteins into their correct shape, which some cannot do by themselves. All proteins are produced in the watery interior of the bacterial cell and have to be water soluble; otherwise, they would rapidly clot and get stuck. But eventually, many TTSS proteins have to function while they are embedded into the membrane, and that is a lipid environment. Proteins that are soluble in water do not like lipids, and the other way round is true as well. To avoid problems, there are folding proteins, called *chaperones*, that "embrace" a protein to lock it in one shape, making it more water soluble, only to let it spring into its lipid-loving (lipophilic) shape upon release, which they do when the membrane is in close vicinity.

All these proteins, with their various functions, work together to produce a functional TTSS. The teamwork is so well coordinated that one can only wonder at the "intelligence" behind this. But all individual components use tricks that evolution had come up with before. Genes have, over evolutionary time, been altered, misused, and combined to produce an impressive, unique, tailor-made combination that makes such a complex final product as a bacterial injection needle. Evolution is powerful, but there is no conductor to lead the orchestra. In the cacophony that evolution produces, some harmonious chords stick when they happen to be useful.

The proteins that make up the TTSS hollow filament resemble those that build the flagellar "tail," which was introduced in the previous chapter. In both cases, many copies of one type of protein form a tube. The assembly machinery that sits in the bacterial membrane to produce a TTSS also mimics the assembly factory that produces flagella. It forms the basis that connects the filament with the bacterial body and consists of many different proteins. Although bacterial TTSS and flagella resemble each other, the two are not the same. Bacteria cannot swim with a TTSS, the filament of which is more rigid than that of a flagellum and does not rotate. And flagella do not act like an injection needle. But most bacteria that produce a TTSS also have flagella. The cell sorts out those proteins that must be incorporated into the TTSS and those that are targeted for the flagella, which is not a small task for a "simple" cell.

What would a bacterium need an injection needle for? That depends on the bacteria producing them, but in most known cases TTSS are produced by pathogenic bacteria, which use them to cause harm to their host. Bacteria use their TTSS to inject proteins (what else?) into a host cell. Every cell is surrounded by a lipid membrane, which makes host cells as impenetrable for bacteria as our skin is for a drop of vaccine. Stick a needle through, however, and in the stuff goes. A TTSS somehow (we do not exactly know how) "senses" the right type of host cell, which it first gives a gentle poke only, to then push through its membrane. The proteins that the bacteria will next inject into the host cell are called *effectors*. That rather friendly term is used for proteins that do quite nasty things to the cells that involuntarily receive them. In many cases, effector proteins force the host cell to start eating the bacteria, which the latter miraculously survive. Now they are where they want to be: inside a host cell, comfortable and safe, where they can multiply and feast on the nutrients that make up the poor cell, while immune cells can no longer reach them. The bacteria grow in number until the cell dies. Then, repeating "knock-knock, poke in, eat me" with their TTSS, they can rapidly spread through the host's tissue. This is what typically happens with an infection by *Salmonella* or *Shigella* species, Gram-negative bacteria that play havoc with your intestinal lining. The result is diarrhea, together with pain, fever, and other symptoms. Fortunately, eventually the immune system will take a gain and the infection will get under control, but fighting an infection takes a toll on the patient's general condition.

The beauty of a TTSS, at least for a geneticist, is that all the genes needed to build it are positioned at one location on the bacterial DNA. This seems logical. It proves handy to put the cutlery in one cupboard and all clothes in another, rather than to store things at random or in alphabetic order. But bacterial genes are not always ordered along the DNA in what we would consider a smart way. In fact, in many cases, complex structures are produced from multiple genes that are scattered all over the DNA strand, as if some bad-tempered geneticist had cut the DNA in a hundred pieces and glued these back together in deliberate chaos. The genes building a TTSS are an exception, as they are neatly positioned together on one location. The explanation for this is that once these genes were combined closely (not by a geneticist or an intelligent designer but possibly just by chance), this provided a very strong evolutionary advantage.

Consider a poor bacterium that lacks a TTSS, but on a lucky day happens to engulf a piece of DNA from a dead neighbor that contains all genes for a TTSS. Once inside, the DNA starts to produce the complete appendage, with regulation, production, and maintenance proteins all doing their jobs as their genes dictate. This bacterium could start to explore novel niches, injecting all sorts of proteins into cells it happens to encounter, until a jackpot is won: a particular combination of effectors injected into one type of cell provides the bacteria with an advantage over its counterparts that had not received that piece of DNA to produce a TTSS. The selective advantage is obvious: the lucky bacteria multiply successfully in this newly discovered niche, and their offspring will produce a TTSS for days to come. This story can only have a happy end if all the genes needed for the TTSS are huddled together, so they can all be transferred in one go. There are more examples for genes that are located together because in that way the chance of their transfer to a new genetic background is increased. As seen in Chapter 12, this also applies to genes making bacteria resistant to multiple antibiotics.

The scenario depicted above has probably happened a number of times. Several bacterial species have a TTSS that we can still recognize as originally belonging to another species. Some *Salmonella* even experienced an uptake event twice, and they now produce two different types of TTSS. These bacteria can eject different effectors under different conditions. My doctor can inject a combination of vaccines in one shot, using only one needle. Instead, *Salmonella* uses different needles for different effectors, keeping track of which effector must be delivered by which needle and when each needle is needed. In this way, it can deliver exactly the right effector at the right time into the right cell. What a shame that *Salmonella* uses such cleverness to cause such a nasty disease, for which we have not even managed to develop an effective vaccine yet.

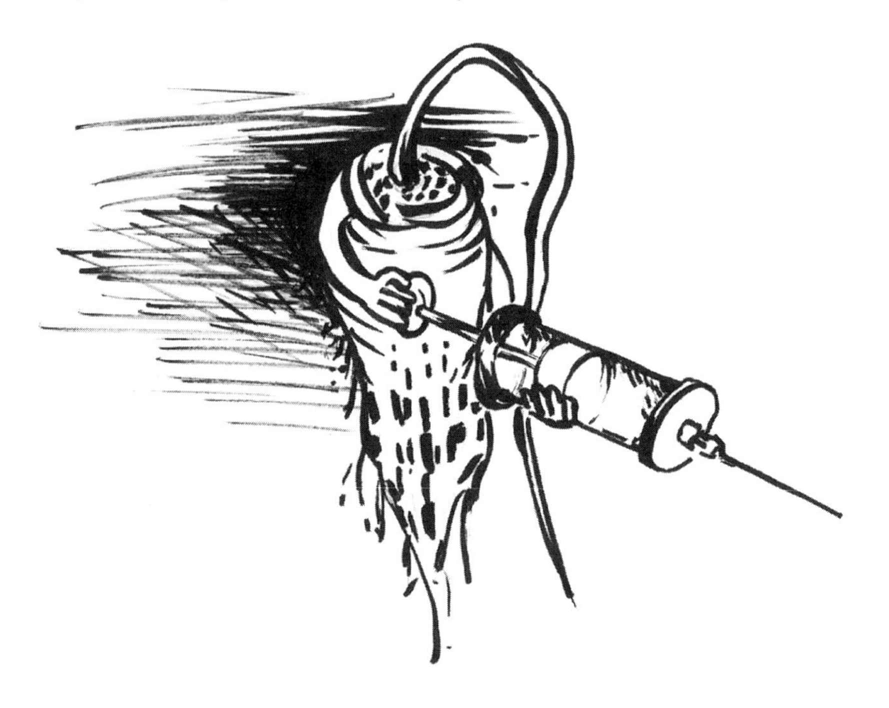

6

Dead or Alive

Complete libraries have been filled with books dealing with the question "what is life?", philosophy's pet subject. Famous books bearing this title have addressed the topic from various viewpoints. Here, the definition and borders of "life" are discussed in a microbiological context, as even microbiologists sometimes have to deal with this complex question. Whether a plant or animal is dead or alive seems easy enough to define, although trees can linger a long time in a state between dying and being dead. The distinction becomes more difficult when dealing with microorganisms. A number of subjects and concepts treated in this chapter will reappear in later chapters, as they are very important to understand the microbial world.

Bacteria are alive unless they are dead; this much is obvious, but how to tell the difference? In practice, we would say a cell is "alive" when it has the ability to multiply, even though it may not be doing so at a given moment. A bacterial spore, for instance, is not multiplying but it still has the potential to do so, and it will change its physiology to start dividing when the environmental conditions permit it to do so. Indeed, multiplication, the ability to produce identical copies of self, is an important component of the definition of life, but it is not the only one, nor is it strictly necessary. A sterile plant or animal cannot produce offspring but it is still alive.

A living thing, be it a single cell or a multicellular plant or animal, must take up energy in the form of food, and change this food into biological mass and waste, through a process called *metabolism* (the word means "change"). Bacteria eat and

Bacteria: The Benign, the Bad, and the Beautiful, First Edition. Trudy M. Wassenaar.

grow, and produce waste in the form of gases, crystals, or soluble waste chemicals. They do so in vast amounts, which resulted in significant global atmospheric changes when the earth was young, as presented in Chapter 3. These metabolic processes consist of multiple steps, each of which is a simple chemical reaction. The combination can be complex, but all processes in a living cell are chemical reactions that follow the basic laws of chemistry and physics.

Reversing the argument of a need for metabolism in order to be alive, a dead cell can be defined as one no longer using energy and no longer converting food into biological material and waste. Metabolism is not only needed for growth but also to maintain self: a living cell will keep its constituents in shape, and any changes caused by diffusion, spontaneous chemical reactions, or external factors will be reversed. Damage that may be caused by chemical reactions such as oxidation, or by radiation, leakage, toxins, and so on will be repaired. The process of maintenance is called *homeostasis*. This requirement for life would push bacterial spores over the edge, as they do not perform homeostasis. Chemical and radiation damage will accumulate over time, until the spore can no longer revitalize. When spores are frozen (as in permafrost soils) this could take a million years, but degradation would be much faster at higher temperatures. So maybe a spore is dead—but as long as it is not permanently damaged it can be restored back to life, as it is in a state of suspended animation.

To maintain homeostasis, bacteria need food, but before turning to their diet, it must be pointed out that the requirement to take up food and produce biological material and waste, results in a need for separating "living matter" from the outside world. A living thing must be surrounded by a border that has to be crossed by food getting in, and waste getting out. Living cells are surrounded by a membrane, a layer of complex lipid molecules that separate "in" from "out." Membranes, sheets of lipid molecules neatly aligned, can form spontaneously from lipid molecules, like small fat drops miraculously appearing and then combining into larger ones on the surface of a plate of soup. Lipid molecules can even spontaneously form three-dimensional spheres. A clever chemist could produce a synthetic "cell" with some carefully chosen molecules inside a lipid sphere that might even be capable of some rudimental metabolism, but this "cell" would still not be alive.

Apart from its inability to reproduce, the synthetic cell of our chemist would never respond to its environment, which is a further criterion for life. Living cells do not always perform exactly the same chemical reactions under all conditions; they respond to signals from their environment and change their chemical reactions accordingly. Without this ability, there is no life. Bacteria respond to their environment: they move toward food and away from repellents, they produce injection needles only when they have to eject effectors or pili when there is a need to attach to a surface, or they change into spores when life gets rough, as we have seen in previous chapters. Chapter 21 will explain how bacteria sense their environments and respond to changes. Those who consider bacteria as not being alive must be ignorant of this capacity.

A final criterion for life is that it evolves over time. A cell whose offspring would always be an exact copy never accumulating any change is, therefore, not alive according to present definitions. It is a hypothetical situation, though, because eventually, mistakes will be made in the process of producing offspring, and such "errors" will accumulate over time. Many changes will be insignificant and some will be deleterious, but once in a while a mutation can be advantageous, for instance, when the resulting cell would be less sensitive to a toxic compound, be better equipped to resist an infecting virus, or better survive a changing environmental condition. The introduced changes were the sole result of sloppy chemistry and their consequences are dictated by chance, but sometimes the living offspring can profit from them, and this selection drives evolution. The combination of mutation (another word meaning change) and selection is the basis of evolution. All living organisms evolve.

In summary, a living cell must be surrounded by a membrane, metabolize nutrients for which it uses energy, respond to its environment, and produce copies of itself when the conditions allow so. Over time, living cells evolve by introducing changes to their DNA. Every cell uses DNA as storage for information, but DNA itself is not alive. It needs to have the cell around it, with its proteins and membrane, in order to express the information so that metabolism can take place, environmental signals can be picked up and translated into a particular response, lipids can be made to form the membrane, and the cell can reproduce (including production of a DNA copy). A cell is not alive without its DNA, and DNA is not alive without the cell. A slight deviation from the rule that all cells contain DNA is the red blood cell (erythrocyte) that loses its nucleus before it starts to work as a transport vehicle of oxygen. Erythrocytes contain vast amounts of hemoglobin, the protein that can bind oxygen to traffic the gas from the lungs to the rest of our body, and we will meet hemoglobin again in a few other chapters. Erythrocytes have lost their nucleus, and thus lack most of their DNA. They can no longer divide, although they exist for approximately 100 days before they are recycled. During this existence, they maintain homeostasis and they would be called alive even in the absence of DNA.

What are the requirements of life for a microbe?

- A living cell must be able to reproduce.
- It must be able to metabolize food to convert this, at the cost of energy, into biological matter and waste.
- A cell must be surrounded by a membrane to separate "inside" from "outside."
- It must be able to maintain homeostasis.
- It must respond to its environment.
- A reproducing organism will evolve over time by accumulation of DNA mutations.

According to the definition used here, a virus is not alive, and most microbiologists agree with this. Nevertheless, viruses meet most of the above-mentioned criteria for life. A virus contains DNA, which bears genetic information (some viruses contain the sister molecule RNA, but that is only a variation of the theme) and this is bordered off from the outside world by a protein coat. Some viruses even have a membrane. Their DNA contains all the information to produce virus copies, although they cannot produce offspring by themselves: all viruses need to be inside a living cell in order to reproduce. However, that is not the main reason why they are considered dead. There are many parasites that cannot replicate by themselves: they live inside a host, sometimes even inside the host's cells, and yet, we call them alive. Independent reproduction is not a necessity for life. Viruses also mutate and evolve, but they are not alive because they do not metabolize, they do not actively respond to their environment, and they do not maintain homeostasis. A virus particle whose DNA is damaged by radiation will not repair this damage. A virus is dead biological matter. However, a virus, although always dead, can still be "killed" in that it can be inactivated, by radiation, desiccation, or chemicals.

The living cell's necessity of food and energy was pointed out as an essential property of life. What do bacteria need for food? Basically, apart from water, which is the solvent inside every cell, they need a source for their carbon, as all living matter is built of this element, as well as nitrogen, oxygen, hydrogen, phosphor, and a number of other elements that are needed in small amounts only. A distinction can be made between bacteria using carbon dioxide (from the air, or dissolved in water) as a carbon source, and those using more complex molecules. Bacteria that use carbon dioxide are collectively described as *autotrophs* as they seem to be "self-feeding" (the literal translation of this term), whereas those that depend on complex molecules are called *heterotrophs*: they need to feed on carbon sources that others produce for them. Another distinction frequently made is whether bacteria need oxygen for growth (or at least survive in its presence), the so-called *aerobes*, or would be killed by this gas, as is the case with the *anaerobes*. Naturally, anaerobes live in environments where oxygen is absent, for instance, in our gut or in oxygen-deprived waters.

A third distinction describes where bacteria get their energy from, as all life depends on an external energy source. This can be light (as in photosynthetic bacteria) for the "phototrophs," or chemical energy in the form of high-energy electrons, which "chemotrophs" can get from high-energy foods, such as hydrogen disulfide, ammonia, or sugars. Moreover, in order to produce biomolecules, the cells require an electron donor: a chemical source of the electrons that are needed to produce complex compounds. Depending on this electron source, the organisms can be further divided into "chemolithotrophs" (when they receive electrons from inorganic compounds, mostly minerals), "chemoorganotrophs" (when they strip electrons off organic compounds), "photolithotrophs" etc., and these terms can again be combined with the "aerobe/anaerobe" or "autotroph/heterotroph" labels. Such descriptive combinations are favored by examiners as much as they are hated by students. It illustrates the fundamental urge of scientists to classify their observations: they put things into descriptive mental boxes in the hope this will increase understanding. (For some students, this does not work, and confusion is only increased by the labels on the boxes.) Although in bacteriology most texts state the energy and food requirements of the subject of investigation using these labels, they are not used in these chapters.

Many organisms live inside other organisms, and frequently their life depends on their host. Such dependence applies both to parasites and to symbionts. A parasitic *Plasmodium* cell (a protozoa causing malaria) cannot live outside its host but it is nevertheless a living organism. This and other parasites live at the expense of their host. Some pathogenic bacteria could be considered parasitic, at least those that can only multiply inside their diseased host, but the term "parasite" is not used for bacteria. Symbiotic organisms are composed of two living symbionts that would both die without the other; they live both at each other's expense as well as to their mutual benefit. An example of a symbiotic organism from the sea is coral, which is formed by colonies of polyps living in symbiosis with algae. Lichens, living on land, are composed of fungi and algae. In these symbiotic life forms, the one component cannot live without the other, but together they are both alive. These are examples of pairs of eukaryotes that form symbionts, but bacteria can live as symbionts, too, and when they do they usually live inside another cell, in most cases inside eukaryotic cells. *Endosymbionts*, as these are called, live inside a host cell and they would die without their host. Again, endosymbiotic bacteria can live as parasites (at the expense of their host) or they can benefit their host. Since they live in a comfortable niche, inside other cells that offer them a home, endosymbionts have frequently lost a number of metabolic processes, as there is no need to produce the compounds their host cell provides them with.

The endosymbionts, bacteria living inside other cells, take us closer to the border of life and nonlife; the term *death* is deliberately avoided here. Before we reach this difficult zone, we must first consider a specialized compartment of a eukaryotic cell that is obviously not a living organism, but has once been alive. The mitochondria have already been introduced as organelles inside eukaryotic cells, and as we will see, they must once have been alive. Mitochondria provide eukaryotic cells with energy carriers. Their name means "threaded granule" and describes their looks under a microscope. They function like little power plants within a eukaryotic cell. Every living cell needs energy, and all cells use the same energy carrier: a molecule called *ATP*. This molecule contains a high amount of chemical energy, which can be used to drive chemical reactions that would otherwise not take place. ATP is the general "fuel" of a cell, and mitochondria produce ATP for the eukaryotic cell, at the expense of burning sugar. In a way, mitochondria are mini-cells, as they possess their own membrane and their own DNA, existing inside the eukaryotic cell. Their size equals that of a bacterium and that provides a first hint that they may once have been bacteria. Their DNA is rather rudimentary, as most of their genes have long since disappeared (these genes can now be found in the nucleus of the cell, where the vast majority of eukaryotic DNA resides), but mitochondria have maintained a few genes for themselves. For this DNA, they use a genetic code that is slightly different from that of the nucleus, but it resembles that of bacteria, our second hint toward the origin of mitochondria.

It is believed that mitochondria were once, in an evolutionary past, bacteria living as endosymbionts inside other bacteria. The synergy worked so well that over time these "proto-mitochondria" specialized in energy production and let their host cell take over more and more of their homeostatic processes, until they evolved into what they currently are: organelles of eukaryotes that are no longer living cells. A mitochondrion is not a living entity, but has become part of a living cell. The current idea is that the original symbiosis occurred between a proteobacterium (a member of the Gram-negative bacterial phylum Proteobacteria) that became the mitochondrion, and an archaea as the host in which this proteobacterium lived. This would explain why present-day eukaryotes strikingly resemble archaea in some of their properties. An alternative hypothesis proposes, as the origin of mitochondria, a symbiosis between two eubacteria, but this fails to explain the similarity between eukaryotes and archaea. If the idea is correct that mitochondria evolved from a symbiosis of a proteobacterium and an archaea, leading to the formation of eukaryotes, this fuses the two prokaryotic domains of the tree of life, Archaea and Eubacteria (see Chapter 2) with the third domain, that of the Eukarya.

The trick of endosymbionts evolving into organelles has occurred repeatedly in biological history and it may be an ongoing process even today. Chloroplasts, to give another important example, have evolved via this scenario, too. These are the organelles in plant cells (animal cells do not have chloroplasts) in which photosynthesis takes place. Photosynthesis will be explained in detail in Chapter 13. Again, chloroplasts have their own membrane and their own DNA with fewer than 200 genes, and, like mitochondria, they probably originated from endosymbiotic bacteria. This time, their ancestors may have resembled present-day Cyanobacteria living in an early animal host cell. Since plant cells have mitochondria, too, it is most likely that the ancestral plant cell evolved from an ancestral animal cell, and not the other way round.

Finally we approach the fuzzy border of life and nonlife in microbiology. Bacteria that live as endosymbionts today are still considered living cells. On the other hand, both mitochondria and chloroplasts are recognized as organelles, specialized compartments of living cells, that themselves are not alive but started off as endosymbionts in a distant past. Endosymbionts (alive) and organelles (not alive by themselves) are two extremes of a scale with a gray zone in the middle where cells (or organelles, depending on the view point) exist in an endosymbiotic relationship with their host cell. They have bacterial DNA, but lack such a high number of typical bacterial genes that it can be argued they are no longer living. For a number of endosymbionts of plant-feeding insects that have been well characterized it is unclear whether they are alive or not. For instance, the psyllids, also known as jumping plant lice, contain compartments in their cells that can be recognized as Proteobacteria from the DNA they contain. But the genomes of these Proteobacteria are so small and incomplete that they can hardly be called bacteria.

They exist at the border of life and nonlife. Sometimes, science lacks the proper descriptive boxes to sort the variation that Nature provides.

Terminology explained

- In contrast to Eukaryotes, Bacteria and Archaea that are all living cells, **viruses** are not alive but parasitize on either prokaryotes or eukaryotes. They consist of a protein or membrane coat surrounding their DNA or RNA, on which their genes reside. Viruses can only multiply by hijacking the reproduction machinery of the cells they infect. Viruses do not maintain homeostasis and do not respond to their environment.
- Symbionts are composed of two different organisms, both of which are dependent on the other for life. **Endosymbionts** are bacteria that live inside other cells and depend on their host for life, which in most cases is a eukaryotic cell. They can be beneficial or parasitic to their host. In the latter case, there is no symbiosis and such bacteria are often described as "obligate intracellular pathogens."
- **Mitochondria** are organelles inside eukaryotic cells that produce the energy-carrier ATP for the cell. They still contain a few genes and are surrounded by a membrane. Mitochondria are the result of a bacterial ancestor (belonging to the phylum Proteobacteria) living in what probably was an archaean cell. They have lost so many of their genes that they are no longer considered living cells.
- **Chloroplasts** are organelles residing inside plant and algal cells that perform photosynthesis to produce sugar from carbon dioxide with light as an energy source. Chloroplasts once were endosymbiotic Cyanobacteria living inside what probably were animal cells. Like mitochondria, chloroplasts are not considered living cells.

7

Nobel Works

When the Swedish Alfred Nobel died in 1896 at the age of 63, he left a fortune, which he had earned with nitroglycerin. He was not the inventor of the explosive, but he made it practically usable in the form of dynamite, which he patented under that name in 1867. After his death, his will dictated that the interest from his fortune be used for prizes in Physics, Chemistry, Physiology or Medicine, Literature, and Peace, much to the dismay of his family. After the fights had been sorted out, the first Nobel Prizes were awarded in 1901 and every year since, only to be interrupted by the two World Wars. A prize in Economic Sciences in memory of Alfred Nobel was added by the Swedish National bank in 1968. Of the 100 prizes awarded up to 2009 in Physiology and Medicine, 20 were given to microbiologists. Who were these great scientists who "made the most important discovery within the domain of physiology or medicine" (the wording Alfred Nobel used in his will), and what did they discover while studying microorganisms? By briefly describing their achievements, we can follow the development of microbiology over the past century in a nutshell.

The very first Nobel Prize in Medicine was awarded for the discovery of diphtheria antitoxin, to Emil Von Behring (1854–1917) from Germany. He started his career as an assistant to the famous Robert Koch, and together they worked on diphtheria. This is a respiratory illness caused by Gram-positive *Corynebacterium diphtheriae* (it belongs to the phylum Actinobacteria) that at that time caused 50,000 deaths annually in Germany alone, most of them children. The disease is spread

Bacteria: The Benign, the Bad, and the Beautiful, First Edition. Trudy M. Wassenaar.

by aerosols. Once inside the body, the culprit bacteria produce a toxin that actually kills the lung cells of the patient. Myocarditis (inflammation of the heart muscle) is a frequent and dangerous complication of diphtheria. From deliberately infected laboratory rats, Von Behring isolated blood serum that could specifically inactivate the diphtheria toxin, and this provided a treatment for humans. The blood and its serum (the fluid that is left after removal of blood cells) contained antibodies: the proteins that the rat immune cells had produced in response to the experimental infection. These antibodies bind to diphtheria toxin which inactivates it completely. Serum therapy came into use once its production could be scaled up and was later replaced by vaccination. Vaccination, as was explained in Chapter 5, results in the body's own production of antibodies, which will protect against disease for years, often life long, while serum can only treat an acute infection and does not provide lasting protection. Thanks to vaccination, diphtheria is now uncommon in developed countries.

A year later, Ronald Ross (United Kingdom, 1857–1932) received the Medicine prize for his work on malaria, but since that is caused by a unicellular eukaryote and not by bacteria, we continue with Robert Koch (Germany, 1843–1910) who received the Nobel Prize in 1905 for his work on tuberculosis. His Nobel lecture still makes interesting reading. Koch had isolated *Mycobacterium tuberculosis*, a bacterium that is neither Gram positive nor Gram negative, as it will not stain at all. It grows so slowly that its isolation was a masterpiece of bacteriology in those days; culturing *Mycobacterium* can still be problematic in modern times. Koch recognized droplets as the source of infection for tuberculosis. He propagated that the chain of infection could be broken by isolating patients with open tuberculosis (when the lung is infected and the patient coughs up the bacteria in sputum) and by improved hygiene and better health conditions. It took a relatively long time until tuberculosis declined, but eventually it did, by a combination of improved living conditions, better hygiene, vaccination programs, and antibiotic treatment. Tuberculosis became a disease of the poor. Sadly, tuberculosis is back as a major killer, mainly in HIV-infected patients in developing countries. In some places, the bacteria have become resistant to a vast number of antibiotics, which further hampers treatment. Koch's hope that "If the work goes on in this powerful way, then the victory must be won," which he expressed at the end of his Nobel lecture, has not yet come true.

His colleague Paul Ehrlich (Germany) was recognized in 1908 for his work on immunity, together with Ilya I. Mechnikov. Immunology is the scientific field that explores how the immune system works: the natural defenses that animals, including humans, have against infectious diseases. Immunology and microbiology were, and still are, closely related research fields. Ehrlich's contributions in both subjects are phenomenal. His industrious life included work on hemolysins (toxins that destroy red blood cells), scaling up production of the antidiphtheria serum, and the development of drugs. At that time, drugs were urgently needed to fight bacteria and parasites, and Ehrlich tried to identify drugs against *Trypanosoma* (a parasite causing sleeping sickness) and syphilis (caused by spiral-shaped Gram-negative *Treponema pallidum* bacteria that belong to a phylum called Spirochetes). His attempts were successful, and he discovered effective treatments against each. Later he became interested in cancer. Ehrlich, whose name lives on in *Ehrlichia* (a type of bacteria that lives as an endosymbiont), would have been thrilled by the idea that some cancers are caused by viral or bacterial infections, as we now know.

In 1928, Charles Nicolle (1866–1936) from France was awarded for his attempts to fight another big bacterial killer: for his work on typhus. This disease, which caused millions of deaths during and after the First World War, is caused by Proteobacteria that live in the intestine of human lice. The bacteria, called *Rickettsia prowazekii*, live intracellularly, just as *Ehrlichia* do. They have a very small genome, lacking many genes so that they cannot divide independently. Their name honors H. Ricketts and S. Prowazek, the two discoverers who both died from typhus as a result of their research. Humans get typhus after a bite from an infected louse or from skin contact with the feces of these insects, which was recognized by Nicolle during his years in Tunis (Tunisia). He worked in a large hospital, and his curious nature drew his attention to the fact that the clothes of typhus patients admitted to the hospital were infective. He recognized that laundry staff frequently became infected, but once bathed and changed, patients were no longer a source of contamination. This set him on the right track to identify the source of typhus as being in the lice feces. His biography reads like a textbook in Medical Microbiology. He introduced a vaccine against brucellosis, caused by various *Brucella* species (Proteobacteria) that are spread from farm animals to man. He was the first to isolate the influenza virus and worked on a number of other viral and bacterial diseases. He recognized that a host can be infected by a disease-causing agent without actually getting ill, nevertheless being able to infect other persons. His Nobel Prize was truly deserved.

In 1939, the Nobel Prize was given to the discovery of an antibiotic. The honor went to Gerhard Domagk (Germany, 1895–1964), who had experimented with Prontosil (an antibiotic from the family of sulphonamides) on his daughter when she was seriously ill from a streptococcal infection. *Streptococcus* bacteria, or "streps" as they are informally known, are Gram-positive bacteria belonging to the phylum Firmicutes. Domagk had already discovered that the red Prontosil compound protected laboratory mice against infection but could not be sure whether it would cure humans when he tried it on his sick child. As a result of the bold experiment, his daughter survived, but he did not disclose the success of this human trial (which would now be considered unethical as he did not know if the used dose was safe for human use) before the drug had been tested on many other patients.

Six years later, Alexander Fleming (United Kingdom), Ernst Chain (Germany/United Kingdom), and Howard Florey (Australia/United Kingdom) shared the prize for the discovery of penicillin. Of the three, Fleming (1881–1955) is best known. The story how he discovered this antibiotic has been told many times. An agar plate of *Staphylococcus* (Gram-positive Firmicutes) had accidentally been left open, and a blue-green mold (probably *Penicillium notatum*) had started to grow, the result of a tiny mould spore that had dropped out of the air onto the nutritional agar. Surrounding the mould growth appeared a circle where the bacteria would not grow. Fleming concluded that the mould must have produced a compound with antibacterial activity, which he called *penicillin*, and that the mould secreted this into the media. He established that this antibacterial activity of penicillin was retained even when diluted 800 times. Chain subsequently demonstrated how penicillin (an antibiotic of the beta-lactam family) destroys bacteria: it inhibits the formation of the cell wall—meaning Gram-positive bacteria can no longer grow (more about this subject is explained in Chapter 12). Florey provided evidence that the drug did not harm animals or humans. The three had thus discovered penicillin, its mechanism of action, and its safety to be used as a drug, which earned them their Nobel Prize. They also shared an interest in lysozyme, an enzyme Fleming had discovered. It breaks down other proteins and is produced in our eyes and saliva to destroy bacteria, as part of our innate immune defenses.

The dominance shifted from Europe to the United States, and in 1951, Max Theiler (South Africa/United States) received the Nobel Prize for his work on Yellow fever, which is caused by a virus. He demonstrated that the disease, which in the 1920s was causing large outbreaks in the United States, could be studied in mice. It did not prevent him from catching the disease himself, but he survived. The availability of a cheap animal model assisted in the development of an effective vaccine. Nowadays, the disease, which is transmitted by mosquitoes, is no longer endemic in the United States but still lingers in tropical Africa and South America.

A year later, Selman Waksman (Ukraine/United States) received the Prize of Medicine for his discovery of streptomycin, the first effective antibiotic against tuberculosis. His interest was broad: he worked on soil bacteria, sulfur oxidation, decomposition of plant material and the formation of humus, marine bacteria, and *Actinomycetes*, a group of bacteria belonging to the Actinobacteria that produce antibiotics. Waksman discovered several more antibiotics during his long career.

The discovery of Poliomyelitis virus that left so many children disabled with polio earned John Enders, Thomas Weller, and Frederick Robbins (all United States) the Prize in 1954. Poliomyelitis virus can infect nerve cells, resulting in lifelong disability. The three researchers discovered how this virus could be grown in tissue cultures, which was a major step forward in studying the virus. Since a virus cannot replicate independently, it cannot multiply in the laboratory unless you find the right cells that can produce virus particles in sufficient amounts. An effective vaccine against polio was developed in 1955, and thanks to vaccination the disease disappeared from most countries. In 1988, the World Health Organization (WHO) set the goal to eradicate polio worldwide, but this has not yet been accomplished. The only disease that has been completely eradicated by vaccination so far remains smallpox (caused by the Variola virus), which was declared accomplished in 1979, after the last natural case had occurred in 1977.

In 1958, Joshua Lederberg (United States) was lauded for his work on bacterial genetics. He shared his Prize with George Beadle and Edward Tatum, who worked, among other things, with fruit flies. Nowadays, it is generally known that DNA encodes information in the form of genes, but Lederberg did valuable work to prove that genes were units of genetic information, years before the structure of DNA was resolved. The latter effort earned Francis Crick (United Kingdom), James Watson (United States), and Maurice Wilkins (United Kingdom/New Zealand) the Nobel Prize in 1962. These scientists would not necessarily call themselves microbiologists, but their names cannot fail here. After all, bacteria depend on DNA just like all other life forms, and microbiological research has provided many insights into how DNA functions in the cell.

Three virologists from France, François Jacob, André Lwoff, and Jacques Monod, received the Prize in 1965 for their work on genetic control of enzyme and virus synthesis. Slowly, the fundaments of genetics became visible, and discoveries in this field continued to hail praise. In 1968, closing the gap between proteins and genes by deciphering the genetic code was awarded with a Nobel Prize for Robert Holley, Har Khorana, and Marshal Nirenberg (all United States). At first, it was thought that the genetic code would be universal in all living things. The code translates the sequence of DNA into protein. DNA is built of four bases, and proteins have 21 amino acids as their building blocks. The genetic code of DNA is hidden in the order of the bases, three at a time, as each triplet codes for one amino acid. Now we know that some bacteria use slightly different codes for a few amino acids or for the signal to tell the cell when a protein is complete. Although the genetic code is generally conserved, it is not completely universal. This is how it was discovered that mitochondria and chloroplasts must once have been bacteria: they use a bacterial genetic code slightly different from that of the cell that harbors them.

That bacteria too can suffer from viral infection was described by Max Delbrück, Alfred Hershey, and Salvador Luria (all United States), who jointly received the Prize in 1969 for this discovery. Viruses attacking bacteria are known as *bacteriophages* (literally "bacteria eaters"), as they seem to eat away the bacterial growth on an agar plate. Phages, as they are known for short, have since been routinely used by geneticists as a result of the work from these three researchers. Phages are viruses, so they are not alive, as was explained in the previous chapter, but they are plentiful. It has been estimated that there are at least 10 times more viruses in the ocean than bacteria, and most of these will be bacteriophages, parasitizing on marine bacteria.

Bacteria usually cause disease soon after they enter a body—but viruses can go under cover and hide for a long time. The discovery of chronic Hepatitis B virus infection and its high prevalence in the population earned Baruch Blumberg the Nobel Prize in 1976, together with Charleton Gajdusek (both United States). This was a bit of a strange combination, as Gajdusek did not study viruses. He worked on kuru, a disease causing mental retardation and early death in Neolithic people living in New Guinea. Gajdusek realized that a similarity exists between kuru and scrapie, a disease in sheep. We now know that kuru was caused by ritual cannibalism which transmitted a "prion."

The puzzle of kuru and scrapie was eventually solved by Stanley Prusiner (United States), who discovered prions as an infectious agent. He received the Nobel Prize for this in 1997. Prions are not alive (just like viruses) and do not even contain genetic material. They consist completely of one type of protein, which is improperly folded. A protein, which consists of a chain of amino acids linked together, can perform its function only when this chain is folded in its proper three-dimensional shape. Changes in this shape will cause changes in its properties. When you beat egg white (which is mainly made of protein) long enough, it turns from transparent and viscous to a solid white mass. This happens because the beating has changed the three-dimensional shape of the proteins, and because of that, they glue together: they have obtained a new property that they did not have when they were in their correct three-dimensional shape. Prusiner discovered that prions are wrongly folded proteins, but the real breakthrough insight was that they could be infectious. Prion proteins force correctly folded proteins of their own kind to refold into the incorrect shape, so that they become infectious prions too. He coined the term prion for such malicious, wrongly folded proteins. Prions are the basis for BSE, or "mad cow disease," which caused a major epidemic in the United Kingdom in the 1990s, infecting thousands of cows. The wrongly folded proteins are so stable that they withstand heating and cooking. Consumption of infected meat caused people to develop a variant of Kreuzfeld's Jacobs syndrome, an incurable, fatal disease. Prions are a new addition to the book of infectious agents, and it was well deserved that two Nobel prizes were dedicated to their discovery.

The next Medicine prize to mention is that of Barry Marshall and Robin Warren (Australia), in 2005, for their discovery of *Helicobacter pylori*. These curved Gram-negative bacteria (Proteobacteria) live in the human stomach, but it took the two discoverers quite some effort to convince other microbiologists of this, who had always thought that the stomach is sterile. It is not, and knowing how versatile bacteria are, it should not have come as a surprise that there are species adapted to living in the acid environment of the stomach. We now know that *H. pylori* is not alone in this environment, as subsequently a wide array of bacteria were found to naturally live in the stomach.

When the first documented cases of a disease we now know as AIDS were discovered, in 1981 in the United States, doctors were puzzled about what caused the unusual outbreak of otherwise rare infections. Patients suffered from *opportunistic infections*, caused by bacteria that can only cause disease when the immune system of the individual is severely impaired. As these individuals had been healthy before then, they must have acquired the condition. The term "Acquired Immune Deficiency Syndrome," AIDS for short, was used to describe the disease, and an infectious cause was presumed. Indeed, in 1983, the virus that causes AIDS was isolated in France, by Luc Montagnier, and in 1984 in the United States, by Robert Gallo. A dispute broke out who was the true discoverer, as both had revealed different pieces of the puzzle, and for a while, one could not even agree on a name for the virus, until HIV, for Human Immune deficiency Virus, won. Although the disease is now frequently called *HIV/AIDS*, the two terms are not synonymous. HIV describes the virus, whereas AIDS is the acronym for the disease. Some individuals are unlucky enough to have become infected, but are lucky in that they do not develop the disease, which will be further explored in Chapter 22. In that case, HIV is not accompanied by AIDS. In 2008, Montagnier and his colleague Francoise Barré-Sinoussi received the Nobel Prize for their discovery, much to the dismay of Gallo. Instead of with him, they shared the Prize with Harald zur Hausen (Germany), who discovered that human Papilloma viruses can cause cervical cancer. Viruses, however, will play only a minor role in the remaining chapters, as there is enough ground to cover with bacteria.

Sometimes it is hard to say whether research was done in microbiology, immunology, or biochemistry, as these fields partly overlap. This summary is not complete, as a number of names were omitted here, which could have been included. A few more Nobel laureates are mentioned in other chapters. It is a good sign that researchers do not stick to scientific borders. Science flourishes by cross-fertilization of ideas from various research fields. There are still many scientific questions that need to be answered, and one can only hope that microbiology will remain to attract the attention of the Nobel Committee.

Bacterial phyla and species introduced so far

- *Salmonella*, *Shigella* and *E. coli*, all of which can cause diarrhea, are members of the phylum **Proteobacteria**. Only some strains of *E. coli* cause disease, while other strains are commensals. Most people have *E. coli* in their gut, but it is by no means the most abundant gut bacterial species. *E. coli* is the guinea pig and working horse of microbiologists. The strains used in the laboratory are crippled, in that they survive only in the friendly conditions microbiologists provide to them. Other bacteria belonging to the Proteobacteria are *Vibrio cholerae* (the causative of cholera), the social *Myxobacter* (that were named after the genus *Myxobacter*), *Brucella* species, *Helicobacter pylori* that lives in the stomach, and the endosymbionts *Rickettsia prowazekii* (causing typhus) and *Ehrlichia* species, which can cause diseases spread by ticks.
- Examples from the phylum **Firmicutes** are *Clostridium tetani*, the soil bacteria causing tetanus, *Bacillus* species, of which spore-forming *Bacillus anthracis* (causing anthrax) is a scary example, and *Staphylococcus*, which assisted in the discovery of penicillin. Not all Firmicutes cause disease, and the Firmicutes is one of the two dominant phyla in the human gut.
- The phylum **Cyanobacteria** has been mentioned without presenting specific genera or species. One of the genera is *Cyanobacterium*, which named its phylum. Others build stromatolites and produce vast amounts of oxygen, which is released into the atmosphere. *Synechocystis* species move along surfaces with the use of pili. Cyanobacteria are the most likely ancestors of chloroplasts in plant cells.
- **Actinobacteria** form another phylum, to which *Corynebacterium diphtheriae* belongs, causing diphtheria. The *Actinomycetes*, frequent producers of antibiotics, and *Mycobacterium tuberculosis* also belong to this phylum.
- The phylum **Spirochetes** was exemplified by *Treponema pallidum*, the causative agent of syphilis.

8

Bacteria and Insects

Most pathogenic bacteria mentioned so far were able to cause disease in humans, but in the bacterial world, humans are only of minor importance or abundance. Of all terrestrial animals on earth, insects represent the most numerous species. Over half of all described animal species are insects, and it is of interest to see how these cope with bacteria, which they meet all the time. Moreover, it is time to turn to some positive properties of bacteria, to counter the impression that they are all bad. Examples of benign bacteria are those that digest the food that animals eat. Insects, like any other animal, depend on bacteria to digest their food. This important task is not carried out by one bacterial species, but by a wide variety of species that simultaneously live in the gut. Such a living community is collectively described as a *microflora*. Similar to our own gut bacteria, the insect's gut microflora produces certain vitamins and amino acids for its host, without which the insect would die.

Termites, for instance, infamous for their appetite for wood, depend on their gut bacteria to feed on this food that is indigestible by all animals. The difficulty in digesting wood is caused by cellulose, which every plant produces as part of their cell walls. Raw cotton is nearly pure cellulose, and about half of the weight of wood is cellulose. Cellulose is composed of long chains of sugar entities, but the way these are connected makes the chains very strong, both mechanically and chemically. Few organisms are able to degrade cellulose into edible sugar, but bacteria (and fungi) have developed the skills to do so. These organisms are responsible for the natural degradation of dead wood, and some of these bacteria have discovered the gut of termites as a comfortable home. In return, the termite

Bacteria: The Benign, the Bad, and the Beautiful, First Edition. Trudy M. Wassenaar.

has learned to live on wood. Since all plant matter contains cellulose, herbivorous animals must have at least some cellulose-degrading bacteria in their gut, or else the cellulose remains undigested. Humans cannot digest it, and neither can our gut bacteria. The fraction of fiber that is an essential part of a healthy diet is mostly indigestible cellulose, whose beneficial effect is that it stimulates gut movement.

The food of an insect dictates what bacteria would live in their gut, or the other way round, as cause and effect can be debated here. Blood-sucking insects live on a diet that is rich in proteins but relatively poor in other nutrients, such as sugars or certain vitamins, for which they rely on their gut bacteria. This is illustrated by the tsetse flies, whose life depends on various bacteria. A lot of research is done on the tsetse fly (*Glossina* species living in the tropics), because their bites can transfer sleeping sickness when the flies are infected with *Trypanosoma* (a protozoa). Sleeping sickness is one of the major tropical diseases that can and sometimes does take a heavy toll on the local human population. Research on tsetse flies has shown that their gut cells contain Gram-negative Proteobacteria called *Wigglesworthia glossinidia*. These bacteria carry over 60 genes that are specialized to produce nutrients for their insect host. In response, the bacteria have lost the ability to carry out some crucial life-saving tasks that the fly does for them. Notably, they no longer possess the gene whose product starts replication, the production of a new DNA strand, which is the beginning of cell division. Instead of producing the protein to start replication themselves, the *Wigglesworthia* cells obtain this protein from their tsetse host, so that the host cells decide when the bacteria can multiply. This makes *W. glossinidia* completely dependent on its host, which also would not like to lose its bacteria, as it would then no longer be able to grow sufficiently in order to reproduce. In addition, tsetse flies carry a second symbiotic bacterial species, *Sodalis glossinidius*, in many of their cells. This endosymbiont (another Proteobacterium) has also lost many of its genes, but it has maintained the ability to build a Type Three Secretion System (see Chapter 5), which it needs to pass from mother fly to the eggs, from which it spreads as the insect develops. It is a rare example of a TTSS that is not involved in disease-causing activities, as the tsetse fly does not suffer from the presence of *S. glossinidius*.

Not only digestion of food but also waste recycling is a "housekeeping" task, and even this can be taken over by bacteria living in insects. Most aphids (such as the green pea aphid, *Acyrthosiphon pisum*) carry Gram-negative *Buchnera aphidicola* in cells that are specialized to harbor these Proteobacteria. As in the previous examples, *Buchnera* has a very small genome lacking so many genes that the bacteria would not survive outside the insect cell, but the aphid cannot grow properly or reproduce without its bacteria, either. It needs the bacteria for recycling its nucleotides, the building blocks of DNA. In most living cells, nucleotides are being constantly recycled so that they do not have to be freshly made all the time, but aphids lack an enzyme needed for this recycling program. Their *Buchnera* take up the half-processed nucleotides from their host and perform the steps for which the aphid has lost the necessary genes. In return, the insect provides *Buchnera* with many compounds that it cannot produce with its small genome.

The effects that bacteria can have on their insect host keep surprising scientists. When *Drosophila*, or common fruit flies, were given a diet that contained either sweet molasses or plain starch, the flies developed a preference to mate with partners that had shared the same food. This sexual preference persisted over generations, but it could be destroyed by antibiotic treatment. Apparently, the cause of this sexual preference must be sought in the bacterial microflora living in the fly's gut, which changes with their diet. The implication of this observation is that bacteria can indirectly influence the formation of novel species. If such a segregation would continue over time, eventually two populations of flies would result, which not only dislike each other but also are no longer able to reproduce, with two new species as the result.

Some bacteria have truly mastered ruling the insects they depend on, and the presented examples are from the phylum Proteobacteria. The genus *Wolbachia* contains many species that live inside insect cells, and they are not choosy: it is estimated that three-fourths of all insects can become infected. In some cases, *Wolbachia* and their host both gain from living together, but there are examples where one can only pity the insect, especially if it happens to be male. When male larvae of some insect species are infected by *Wolbachia*, they are killed, whereas female larvae survive, so that the sex ratio of the insect population is changed. Alternatively, infected males develop as females. This sexist lifestyle is not specific to *Wolbachia*: species of *Rickettsia* (another bacterial genus with an intracellular life) kill male ladybird beetles. The population of *Trichogramma* wasps, tiny predators that feed on eggs of caterpillars, exists completely of females, for which *Wolbachia* are to blame. Since this obviously poses reproductive problems, the female wasps can, again thanks to *Wolbachia*, reproduce without the help of a male. The bacteria's preference for females is understandable, as they are transmitted from the mother insect to her offspring. From the bacteria's point of view, male insects are useless. Fortunately, male-killing bacteria have so far evolved only in insects.

Insects seem to carry a wide variety of endosymbionts, and some of these were discovered in unconventional ways. After the complete genome sequence of the common fruit fly (*Drosophila melanogaster*), which was the first insect to have been sequenced, had been elucidated, various other *Drosophila* species were subsequently (partially) sequenced. Within the long DNA fragments that were produced during this work, some of the obtained sequences obviously did not belong to the fruit fly, as they were bacterial, so they were discarded as "contamination." Soon, they were salvaged from the trash, as it was realized that they belonged to a novel type of *Wolbachia*. The pieces of DNA were put together to produce the nearly complete genome of what is now called *Wolbachia pipientis*, which lives as an endosymbiont in the fruit fly and, as was later discovered, in other insects. It is the first "wastebin" bacterial genome that was recorded, but it is unlikely to be the last endosymbiont to be discovered this way. It was subsequently discovered that *W. pipientis* plays an important role in river blindness, caused by tiny worms that are transmitted to humans by black flies. The worms harbor *W. pipientis*, and both the bacteria and the worms contribute to the disease. That bacteria living inside worms living inside insects can cause blindness in people sounds like some nightmare Russian doll with a jack-in-the-box inside.

A more common association that people have for "bacteria" and "insects" are diseases that are caused by (biting) insects, and there are an impressive number of these. However, very few of these diseases are actually caused by bacteria, as most are due to either eukaryotes or viruses. Between 300 and 400 species of insects exists that regularly suck human blood, and many of these can transfer disease, in which case the insect is called the *vector* of that disease. Six vector-borne diseases together cause nearly 500 million people around the globe to fall sick each year. These are (in decreasing order of prevalence) malaria, elephantiasis, Dengue fever, river blindness, Chagas' disease, and leishmaniasis. Not one of these diseases is caused by bacteria but they are briefly summarized here.

Malaria, the most prevalent vector-borne disease, is caused by *Plasmodium*, a unicellular protozoan eukaryote. It is transmitted through mosquitoes of the genus *Anopheles*. Infectiologists group pathogenic (disease-causing) eukaryotes as "parasites" to differentiate them from bacteria and viruses, so *Plasmodium* is counted among the parasites. That terminology is a bit confusing, as parasites can be both multicellular worms and unicellular amoebas alike. Moreover, parasitic life styles are not exclusive to eukaryotes, so the term "parasite" will be avoided here. The other major pathogenic protozoa that depend on an insect vector are *Trypanosoma cruzi*, causing Chagas' disease, brought on by the biting *Triatoma* bug, and *Leishmania*, transmitted by sand flies. Elefantiasis (also known as lymphatic filariasis) and river blindness are caused by microscopic worms, and in both cases these worms depend on *Wolbachia* bacteria to be alive. This raises the possibility to treat these diseases with antibiotics. Dengue fever is a viral disease. Elefantiasis and Dengue fever are spread by mosquitoes, whereas river blindness uses black flies as a vector.

Diseases caused by eukaryotic microbes

A number of human diseases that are spread by insects but caused by eukaryotes instead of bacteria have been mentioned.

- **Malaria** is caused by various *Plasmodium* species. It is spread by mosquitoes of the genus *Anopheles* and is restricted to tropical and subtropical climates. Ronald Ross received the Nobel Prize in 1902 for his work on malaria. *Plasmodium* is distantly related to dinoflagellates and still contains the remnant of what once was a chloroplast. Obviously, these intracellular parasites can no longer perform photosynthesis, but in an evolutionary past, their ancestors may have been algae.
- *Trypanosoma* are protozoan parasites causing **sleeping sickness** predominantly in Africa (*Trypanosoma brucei*) or **Chagas' disease** in Central and South America (*T. cruzi*). Sleeping sickness is spread by the tsetse fly, whereas Chagas' disease depends on the *Triatoma* bug. The cells of *Trypanosoma* arc somewhat special, as they contain only one big mitochondrion (most other cells contain multiple smaller ones) that harbors multiple mitochondrial DNA copies in a big entangling structure.
- **Leishmaniasis**, spread by sand flies, is caused by various *Leishmania* species, which are closely related to *Trypanosoma*. Most *Leishmania* cases are transmitted from animals to humans; human-to-human transmission is rare but does exist for some species.
- **Elephantiasis** is caused by parasitic worms (multicellular organisms of various genera) generally called Nematodes, which are transmitted by mosquitoes. The worms carry *Wolbachia* species that they depend on for growth, and these bacteria are strong triggers of the human immune system. Some of the symptoms are thought to be due to an excessive immune response for which *Wolbachia* may be mostly responsible.
- The roundworm *Onchocerca volvulus*, a member of the Nematodes, causes **river blindness**. It depends on biting blackflies (*Simulium* species) for its spread. Patients turn blind as a result of their immune system running wild, which is in part caused by the *W. pipientis* bacteria living as endosymbionts in the worm's cells.

All vector-borne diseases mentioned so far in this chapter are caused by either eukaryotes or viruses, but that is not to say that insects cannot be vectors for bacterial disease. The importance of lice in the transmission of typhus was already pointed out in the previous chapter. The link between fleas and *Yersinia pestis*, the causative agent of plague, is generally known, and ticks are infamous for transferring Lyme disease, an infection caused by *Borrelia burgdorferi*. These spiral-shaped, Gram-negative bacteria (for a change they belong to the phylum Spirochetes) are peculiar, as they have their flagella inside their membrane. When their flagella rotate, they cause the whole bacterial body to change shape, and this wiggling makes them move forward. Lyme disease is named after the town Lyme in Connecticut, United States, where a cluster of cases occurred in the mid-1970s, although the disease had been known for decades. Borreliosis, as it is also known, is the most common tick-spread disease in the northern hemisphere.

Ticks are vectors for a number of other bacterial diseases, a few of which are worth mentioning, ignoring the fact that ticks are not insects but belong to the Arachnids. *Ehrlichia* are another group of bacteria that live inside the cells of ticks or of the hosts on which these ticks feast. Dog owners in the tropics and subtropics should watch out for CME (canine monocytic ehrlichiosis), caused by *Ehrlichia canis*, which can be potentially lethal for their dog. Only one particular type of tick can transfer the disease, and its habitat is restricted to the warmer climates. Thus, the climate requirements of the insect vector restrict the geography of the bacterial disease as well. Once a dog is infected, it will remain so lifelong, even after antibiotic treatment. Apart from dogs, other animals such as sheep, cattle, and horses also frequently suffer from infections they received through tick bites. A cattle disease called heartwater (after the accumulation of fluid around the heart of severely sick animals) is caused by a bacterium previously known as *Cowdria ruminantium*, which is now called *Ehrlichia ruminantium* (note the changed genus name) and used to be exclusive to Eastern Africa. Ticks carrying these bacteria were accidentally imported into the Middle Americas and then into some Southern states of the United States. Humans were probably to blame for the crossing of the Atlantic, but the disease was spread further by migrating birds, notably the Cattle Egret, which can transport infected ticks. Local farmers and veterinarians have to keep a careful eye on their cattle to avoid further spread of the disease.

In all these cases, the insect vector is not harmed by the presence of the pathogen, but insects can also themselves suffer from bacterial infections. Insect pathogens can be very specific or attack a broad range of insects. *Serratia marcescens* (again a Gram-negative Proteobacterium) is not very selective, as it can make a variety of insects sick. When the bacteria are eaten, they upset the insect's gut and invade deeper tissue, not unlike what *Salmonella* can do when it infects us. A wounded insect had better stay away from *Serratia* as well, since this can also enter through wounds and kill its host that way. *S. marcescens* can infect humans, too, but such infections are not very common, although they are much dreaded in intensive care units of hospitals, especially in those for newborn babies. They are an example of an opportunistic pathogen that can cause disease only in individuals with a weak immune system. The bacteria can cause outbreaks in neonatal intensive care units, as the immune system of premature babies is not working fully yet. Such a vulnerable host is an easy victim for *Serratia*, whereas the immune system of a healthy adult would probably eradicate the bacteria before much harm could be done. Insect larvae, the insect equivalent of babies, are also more susceptible to bacterial disease than adult stages. *Pseudomonas entomophila*, whose species name already suggests it "likes" insects, specifically kills fruit fly larvae. The immune system of the insects (yes, insects have an immune system, too) cannot do anything to stop these Gram-negative bacteria, which block food uptake in the larval intestine so that the insect starves to death.

This brings us to biological pest control. When bacteria can kill insects, why not use these natural enemies to combat insects that threaten our crops? There have been examples where this approach worked. Japanese beetle (*Popillia japonica*) is a pest to both crop and ornamental plants in the United States because of its big appetite for leaves, which it nibbles until only a skeleton of leaf nerves remains. Their larvae feed on plant roots, which add to the damage this pest causes. The beetle was imported from Japan, at the beginning of the nineteenth century, where it is far less of a pest, since it has natural enemies in its country of origin. A natural pathogen for the grubs of *Popillia* is *Bacillus popillae*, which causes milky disease—named after the white appearance of the infected grubs. The spores of this Gram-positive Firmicute are produced on a commercial basis to keep the beetle pest under control. However, the beetles must be infected with relatively large numbers of spores to be in serious trouble, which the bacteria can only naturally produce when sufficient grubs are around. This illustrates the problem of using a pathogen that needs its host to multiply. Although its specificity is desirable, as the spores will not be toxic to any other insect, it also reduces its application as a pest control. Once the propagation of grubs and spores has reached critical numbers, Japanese beetles can indeed be kept in check by *B. popillae*, especially when the spores are applied to large areas. This became the first registered microbial control agent commercially available in the United States.

An insect pathogen with a more generalist lifestyle is *Bacillus thuringiensis*. It is commonly present in soil around the world. When insects eat the bacteria, they suffer from intoxication, because the bacteria produce a toxin that disturbs the insect's digestive system. The intoxicated host stops eating and dies of starvation. The toxin, "Bt" for short, is a crystalline protein that remains active even when the bacteria are dead. *B. thuringiensis* is also used for biological pest control. Particular strains of these bacteria are active against certain types of insects, so a farmer has some means to target the biological control agent. In fact, these differences are based on different types of toxins produced by different strains of this diverse species.

From the application of live *B. thuringiensis*, chosen to fit the correct crop and its threatening pests, it was a small step to introduce the gene that codes for the right Bt toxin into the plant itself. The result is a crop plant that has become toxic for the insects that want to feed on it. Such genetically modified crop plants have been developed for maize and cotton, and these now make up a large percentage of the maize and cotton produced worldwide. As with all novel technology, there is opposition, as some people do not like the idea that we meddle with plant genes. Although Bt is not toxic to humans, there are still some risks involved, such as harming benign insects or disturbing an ecological balance. Insects may also become resistant to the toxin, although so far there are no signs of this happening. At present, some countries are more willing to accept products from genetically modified plants than others. It makes life complicated for traders, who have to keep harvests (or their products) separate depending on whether the plant was, or was not, genetically modified.

Life is a matter of "eat or be eaten" as well as "feed and be fed." Flies feeding on our blood happen to be infected with protozoa that make us sick when they feed on us, but the flies themselves are infected with bacteria that provide them with nutrients, whereas the insects feed the bacteria in return. Aphids parasitize on plants, feeding on the plant's sap, but they are again parasitized by bacteria that live inside their cells. Insects eat plants (and need bacteria to digest these for them) that we would like to eat or that we feed to the animals that we farm for food, but in doing so, these insects eat bacteria that kill the insects that would eat our plants. It could make a nice nursery rhyme. We can exclude the bacteria from this sequence of events by genetic manipulation, but we should not forget that none of these modern-time agricultural miracles would have been possible if we had not learned how to manipulate genes. Guess what model system was used to practice on? Bacteria, which manipulate genes, of plants or insects, all the time.

9

Bacterial Toxins

Mrs. White checked the dining room one more time. Everything seemed in perfect order, which was no surprise as she had been meticulously preparing the dining table all day for the guests who were expected that night. She had ironed the tablecloth and napkins, bought fresh flowers of a color matching the cutlery, and rinsed all glasses an extra time, although they had already been spotless when she took them out of the cupboard. In the kitchen, most of the food was ready, too. Nothing special, in the eyes of Mrs. White, but honest, home-cooked food that she was proud to serve. The salad still had to be made, as this had to be fresh, but she could already collect all ingredients to have everything at hand. Down in the basement she took a jar of home-canned carrots, which would add a nice colorful touch to the vegetable salad she was planning. On her way back to the kitchen, she unscrewed the lid and dipped her finger in to take a taste. She always tasted food before serving it, out of habit. It would be a shame if too much, or too little salt had been added, or anything else was not perfect. Mrs. White took pride in her skills as a cook and a hostess.

During preparation of the salad, she changed her mind and replaced the carrots with red pepper, as the color effect would be even stronger. "A good cook can improvise and has to be creative," Mrs. White thought. And she considered herself a good cook. That night, her guests confirmed the opinion she had of herself. The food was delicious and the company thoroughly enjoyed the dinner.

Two days later, Mrs. White felt a bit dizzy and nauseated, which was quite unusual. She could not work out what was wrong with her, but whatever was

Bacteria: The Benign, the Bad, and the Beautiful, First Edition. Trudy M. Wassenaar.

bothering her, it was getting worse very quickly. When walking became difficult, she urgently arranged to see the doctor. By that time her speech had become slurry, and she was rushed to hospital in an emergency ambulance with the suspicion of a stroke. Her muscle weakness progressed rapidly, and by the time the doctor examined her she had to scribble her answers on a note pad, no longer being able to speak. Fortunately, the doctor asked the routine question that is so important in these cases: had she eaten any spoiled or home-canned food lately? The carrots stood out in her memory like red flags. Within minutes, she was on medication against botulism, which turned out to save her life just in time. Had she not changed her mind about the salad, the dinner party could have led to a number of funerals. Mrs. White was lucky to have survived, but her recovery was extremely slow and she had to go through a lot of suffering.

Botulism is not frequently encountered any more, since the habit of canning food at home is now less common. Home canning of vegetables that are low in acid using the water boiling method is strongly discouraged, as botulism is a small, but serious risk. Industrially canned food is carefully checked for *Clostridium botulinum*, the Gram-positive bacterium (it belongs to the phylum Firmicutes) that is the cause of botulism.

As is the case with a number of other infectious diseases, the *C. botulinum* bacteria are not themselves responsible for the disease, but during growth they produce an extremely powerful toxin. Consumption of the toxin is what causes botulism, even if not one live *C. botulinum* bacterium was ingested. The reason canned food poses a risk is that these anaerobic bacteria can only multiply (and produce their toxin) in the absence of oxygen, and exactly those conditions apply inside a food can. Moreover, the bacteria produce spores, like many other Firmicutes do, and these spores survive boiling. Once inside the can, the spores start growing and add the toxin to the food. Cooking would destroy the toxin, but canned food is typically only briefly heated, which is insufficient to inactivate all the toxin. Acid-containing foods, such as fruit, tomatoes, and fermented or pickled food pose no risk, as *C. botulinum* is acid sensitive and cannot stand high salt concentrations.

How does botulinum toxin, the most potent biological toxin known to man, work? The protein, of which one gram could kill a million persons, is a *neurotoxin*. It works by binding to the synapses of the human nerve cells that communicate with muscle cells. Synapses are the communicating ends of a nerve cell, and once the toxin is bound to these, it cuts through an important protein in their membrane, so that they can no longer process nerve signals. The result is muscle weakness, paralysis, and eventually death. After ingestion, the toxin can cross through the cells lining the gut and reach the bloodstream, from which it eventually reaches the nerve cells to bind with high affinity to their synapses. Soon after the mechanism of action of this bacterial toxin was elucidated, it was realized that one could use it to relax undesired muscle tension, when applied locally and dosed carefully. The toxin is now medically used to release spasms and tremors. It is even used cosmetically, where it is known under the trade name Botox. By injecting extremely diluted amounts of the toxin into the skin of the face, local muscles are relaxed, which smoothens out the frown lines of many a celebrity. The toxin works for a few months, after which the injection needs to be repeated.

Organisms related to *C. botulinum* produce toxin too. *Clostridium tetani*, already mentioned in the first chapter, produces a neurotoxin that is closely related to Botox, but instead of causing paralysis, it forces the muscle to contract, again by binding to nerve cell synapses, resulting in the spasms typically seen with tetanus patients. *Clostridium perfringens*, however, produces a toxin that works in a different way. These bacteria are a more common cause of less dangerous food poisoning than their botulism cousins. *C. perfringens* produces a protein that can bore small holes in a membrane (although not in their own membrane). It belongs to the family of so-called *pore-forming toxins*. The importance of an intact membrane has already been stressed, so it is no surprise that bacteria have discovered that membranes are the Achilles heels of a host cell. Perfringolysin, as the toxin produced by *C. perfringens* is called, is a protein that spontaneously binds to the cell's lipid membrane. When multiple protein molecules do this, they spontaneously form a ring with a hole in the middle while sinking into the membrane. This donut-shaped protein complex produces a leak through which electrolytes can leave the cell, which is severe enough to cause cell death. This is one of the arrows *C. perfringens* has to the bow to cause diarrhea, as it kills the cells of the intestinal lining.

C. perfringens also produces another toxin, and with this it can cause another, far more serious disease: gas gangrene. The disease results from *C. perfringens* bacteria growing inside a closed wound (these bacteria cannot stand oxygen) where they play havoc. The toxin responsible for the damage is an enzyme that can split lipid molecules called phospholipids, which are a major constituent of the host cell membranes. Because of its activity, the enzyme is called phospholipase (an enzyme name ending on "-ase" often means something is cleaved). The result of the phospholipase activity in an infected wound is that the tissue literally melts away, with devastating consequences. When gas gangrene develops, the treatment can be amputation of the infected limb, as the massive cell death releases so many waste products that the patient would die from his or her own detritus.

Cell damage by pore-forming toxins is quite a common mechanism for causing disease. The Bt toxin, produced by *Bacillus thuringiensis* (discussed in the previous chapter), also acts by forming pores in the insect's gut cell membrane. Pore-forming toxins have been extensively studied. When such a toxin is added to a solution containing erythrocytes (red blood cells), it shoots holes in them through which the hemoglobin is released. The result is easily visualized in the laboratory, by centrifuging the vial containing the treated blood cells. Whereas centrifugation of intact red blood cells results in a small red pellet with a colorless solution above it, the released hemoglobin from killed erythrocytes remains in the solution even after centrifugation, which now turns bright red. This "hemolysin" test is routinely used to study pore-forming toxins, and it is the reason why many pore-forming toxins are known as *hemolysin*, even though red blood cells may not be their major natural target.

Some dangerous *Escherichia coli* types that can cause serious food poisoning produce a hemolysin. The type of *E. coli* that caused some very large outbreaks related to contaminated beef, which the press nicknamed "hamburger bug," are scientifically known as *type O157*. These *E. coli* O157 cells produce a potent hemolysin, but that is only one component of their pathogenic repertoire. They also possess a TTSS, as well as a toxin that functions similar to Shiga toxin, discussed below. This type of *E. coli* is particularly dangerous in children and infants, as the toxins they produce indirectly result in kidney damage. Kidney failure, or HUS (hemolytic uremic syndrome) is a potentially fatal complication that can occur during an *E. coli* O157 infection. A large O157 outbreak caused by infected meat struck Scotland in 1996, involving 512 cases and killing 22. The way this outbreak progressed followed a pattern frequently observed in disasters: a chain of events took place that individually could easily have been prevented. Each was considered unlikely to occur; nevertheless, they combined to develop into a catastrophe. The outbreak was managed badly, and in hindsight serves as an example of how things can grow out of control. Risk researchers have learned from this and other disasters to anticipate a combination of unusual and unexpected events that in combination can have disastrous consequences.

Even so, outbreaks can not always be prevented. In 2011 a very large outbreak of *E. coli* with exceptionally frequent HUS complications occurred in Germany, involving thousands of cases, and (at a smaller scale), later in France. It happened to be caused by an unusual *E. coli* type, not of the O157 but of the O104 type, and it took quite some time to identify the source. Contaminated bean sprouts eventually were found as the culprit, and knowing this the outbreak could be ended, but more than 800 HUS cases had already occurred, and over 40 people died.

So far, we have dealt with toxins binding to nerve cell synapses or toxins damaging cell membranes via various mechanisms. Another way to cause a cell to die is by forcing it to flush itself empty. *Vibrio cholerae* bacteria produce a toxin that does this. These bacteria infect the intestine, and cause a massive watery diarrhea by means of their toxin. The cholera toxin is a large protein, composed of six parts; it is sometimes described as an AB_5 toxin. Five identical protein B parts form a donut-shaped ring with the sixth, different protein A filling the donut's hole with its tail. As one might expect, the donut binds to cell membranes but this time it does not produce a hole. Instead, upon landing in the membrane, it releases the A component by cutting it loose. This cutting not only releases but also activates the smaller A protein into the actual toxic part that has enzymatic activity. As a result of the donut binding, the host cell starts to "eat" this now toxic enzyme A. Once inside the target cell, the active enzyme A disturbs a critical regulatory process of the cell in which one small molecule is crucial: cyclic AMP. Cyclic AMP is related to ATP, but it does not serve as an energy carrier. The function of cyclic AMP in the cell is that of a guardian that regulates normal channel proteins that are found in every cell membrane. These channels normally allow specific ions to pass, to enter or leave the cell upon demand. Through a chain of events inside the cell, cyclic AMP ensures that the right type of channels are closed or open, depending on the cell's need. But cholera toxin interferes with the amount of cyclic AMP that is produced in the host cell, and as a result all channels flush open. The cellular electrolytes stream out, and the cell dies. The bacteria, which stay in the lumen of the gut, can feast on the released cell content that is now available to them. The massive loss of electrolytes and cell contents is the basis of the voluminous watery diarrhea that is typical of cholera. A cholera patient can lose liters of fluid per day, so that dehydration and loss of electrolytes are the major cause of death.

There are more bacterial toxins that interfere with cellular cyclic AMP levels. Another example is pertussis toxin, another AB_5 toxin, produced by *Bordetella pertussis*, the causative agent of whooping cough. This time, the lung cells are targeted as this pathogen (also a Proteobacterium) infects the respiratory tract. Another example of a toxin manipulating cyclic AMP levels in the host cell is produced by *Bacillus anthracis*, the cause of anthrax (which, by the way, also produces a second toxin that works in a different way). Note, that *Vibrio cholerae, Bordetella pertussis and Bacillus anthracis* are completely nonrelated bacteria, but their toxins function in a similar way. This is a nice example of parallel evolution.

Two more toxins are worth mentioning, as they interfere with yet another essential function in the target cell: protein synthesis. Before the mode of action of these toxins can be discussed, we need to understand how cells produce their proteins, using the genes on their DNA. Bacterial and eukaryotic cells alike make protein from DNA by means of an intermediate molecule, called *messenger RNA*. When a particular protein is needed, its gene becomes active, which means it will be copied into a messenger RNA. These short sister molecules of DNA are made for each gene whose protein is needed in the cell at a given time. Whereas every cell contains all genes of a genome in its DNA, only a few of these genes are used at any given time to produce RNA. This way the same DNA is shared between, say, a liver and a muscle cell but their properties depend on which genes are used to produce messenger RNA. This messenger RNA is subsequently recognized by *ribosomes*, little "machines" that are responsible for the production of protein in all cells. They look a bit like a doll, with a small head sitting on a big belly, and are made up of many proteins and a special kind of RNA. The ribosomes function as little "knitters" since they combine loose amino acids into long strings that become proteins. In order to produce the right kind of protein, ribosomes slide along a molecule of messenger RNA and "read" the genetic code on it. This instructs them which amino acid to add to the growing chain, thus producing a protein. Without active ribosomes there is no protein production.

This is where Shiga toxin hits, the toxin produced by *Shigella dysenteriae*. Shigellosis is another result of food poisoning. The structure of Shiga toxin resembles that of the cyclic AMP-affecting toxins, with a ring of five proteins bound to a further protein: it is another AB_5 example. The A protein again carries out an enzymatic reaction after it enters the target cell, but this time it attacks the cell's ribosome. Shiga toxin prevents the eukaryotic ribosome from doing its job: it can no longer produce protein. Without protein production, all cellular processes eventually come to a stop and the cell dies. A protein very similar to Shiga toxin is produced by the *E. coli* O157 type mentioned above. It is quite possible that one of the two organisms, *S. dysenteriae* and *E. coli* O157, donated the gene for this toxin to the other, although it is not known in which direction this gene transfer took place. Maybe they both received it from yet another, so far unidentified, donor. What we do know is how the gene moved place: it is part of a bacteriophage (a virus parasitizing on bacteria). Bacteriophages are often the vehicles by which genes are transferred from one bacterial species to another.

To continue with toxins that target protein synthesis, we move on to diphtheria toxin, produced by Gram-positive *Corynebacterium diphtheriae* (mentioned in Chapter 7). This toxin interferes with protein synthesis too, although its target is not the ribosome but one of the other proteins needed in the process. The toxin uses a similar enzyme activity as Shiga toxin, and the effect is the same: the cell dies because protein synthesis is halted. *Shigella* bacteria live, and deposit their toxin, in the intestine, whereas *C. diphtheriae* lives in the throat. Again, the two toxins are not genetically related, but they both target protein synthesis, as this is another weak side of the host cell.

It is not out of a morbid interest that microbiologists study bacterial toxins in such great detail. By understanding how exactly a pathogen damages its host, one might be able to design specific drugs to prevent these steps. For instance, bacterial toxins have been used to produce vaccines that specifically inactivate their toxic activity, and this proved a successful strategy to combat a number of diseases. As the Botox example shows, toxins can sometimes be used beneficially, or they are used as tools to investigate fundamental cellular processes. Toxins are frequently used in the laboratory, for instance, to halt protein synthesis of a cellular population under investigation. Shiga toxin is now being investigated as a drug to kill cancer cells. We can make the best of these proteins because bacterial toxins are extremely good at one thing: they are fantastic killers.

Targets of bacterial toxins

Bacteria produce toxins that target sensitive parts or processes of the eukaryotic cell:

Membrane integrity and membrane function

- Pore-forming toxins bore holes in the cellular membrane so that the cell becomes leaky and dies. Examples are various toxins produced by *C. perfringens*, some *E. coli* strains, *B. thuringiensis* or *B. anthracis*. *Staphylococcus aureus* also produces a pore-forming toxin.
- Membranes can also be attacked by enzymes that degrade their lipids, for instance, by the phospholipase produced by *C. perfringens*.
- The membrane can also be forced to malfunction, by opening all its channels, and this is how cholera toxin works. It is an AB_5 toxin whose active component A has to enter the cell to manipulate the cyclic AMP concentration, which is a guardian of channel activity. A less potent toxin of this kind is produced by some *E. coli* strains. Both bacteria received this toxin by means of a phage.

Protein synthesis

- Halting protein synthesis is an effective way to kill a cell. Shiga toxin, produced by *S. dysenteriae* and some *E. coli* strains, inactivates the ribosomes of the intestinal cells after entering these. Shiga toxin is an AB_5 toxin whose active A part makes a small cut in the cell's ribosome. This completely inactivates the ribosome, and this halts protein synthesis, as a consequence of which the cell dies.
- *C. diphtheriae* produces a toxin that halts protein synthesis by manipulating a protein called Elongation factor-2, which is essential for eukaryotic protein synthesis. This kills the cells in the throat, where these bacteria multiply.

Cell function

- A second toxin produced by *B. anthracis* elevates cyclic AMP levels, like cholera toxin does, but the target cells differ and thus the effect: the toxin manipulates cyclic AMP inside immune cells, which start to malfunction so the host can no longer elicit a proper immune response.

- The toxin produced by *B. pertussis*, which infects the lungs, also manipulates cyclic AMP levels in immune cells.
- Botox produced by *C. botulinum* cleaves an important protein on the surface of nerve cell endings, the synapses. The toxin is a zinc-metalloprotease, meaning the enzyme can only cleave its target protein when it is bound to zinc.
- The zinc-metalloprotease toxin of *C. tetani* works similar to that of *C. botulinum*, but results in spasms rather than paralysis. This explains the differences in symptoms between tetanus and botulism.

10

Enzymes

The term "enzyme" has been used several times in previous chapters without an explanation, but what exactly does it mean? Enzymes are proteins that work as a catalyst. This term originated in chemistry: a catalyst is a chemical, or compound, that speeds up a reaction. In a chemical reaction, one or more compounds, called *substrate(s)*, react to produce one or more products. The simplest reaction can be given by the formula A→B, whereby substrate A changes into product B. In a little more complex reaction, A + B→C would represent two substrates combining into one product, whereas A→B + C describes a reaction of a single substrate forming two products. The complexity can increase as in A + B→C + D, and so on. Any of these types of reactions can be speeded up by a catalyst, so that it takes place more rapidly than would be the case without the catalyst. Enzymes are proteins that speed up a chemical reaction in the cell. They do this by binding to the substrate (the component or components of the chemical reaction that needs to react) and facilitating the chemical reaction to produce one or more products. At the end of the reaction, the enzyme lets go of all components and is ready to repeat its job with a new substrate molecule, as it has not undergone any chemical change itself.

The binding of substrate and enzyme is very specific and has been compared to a lock and a key: a wrong key would not fit a lock, but when correct, together they open a door. Similarly, a wrong substrate-enzyme combination would not perform any reaction and, in the worst of cases, would get stuck, but in the right combination they work. How do enzymes work? They might catalyze a reaction by binding two different substrates at the same time, which greatly increases the chance that these

Bacteria: The Benign, the Bad, and the Beautiful, First Edition. Trudy M. Wassenaar.

two molecules meet at all in the cell soup and are able to react. Or the enzyme provides a local chemical environment for the substrate, which enables a reaction that would otherwise not, or only slowly, occur. More complex ways of catalyzing a reaction are also possible, for instance, when energy is simultaneously added to the process, by the splitting of ATP molecules (the energy carriers of a cell). Every time an ATP molecule is split, chemical energy is released, which is how ATP serves as the fuel of a cell. Enzymes can "collect" that energy and transfer it in a steplike procedure to the substrate in question, so that it is sufficiently energetically charged to produce the desired product.

Enzymes are the working horses of living cells. They are responsible for the chemical reactions that together provide all metabolic processes in the cell. Enzymes produce the lipids of the membrane; they link proteins to these lipids, or lipids to phosphate, phosphate to proteins, sugars to lipids or proteins, or splice these, and do a lot more. Enzymes bind to, and thus inactivate, toxic radicals (very reactive chemicals that are produced as a side product of metabolism or as a result of oxygen damage) or inactivate other toxins, including antibiotics. Enzymes produce nucleotides, the building blocks of RNA and DNA, and they build RNA and DNA from these nucleotides. Enzymes also produce amino acids, the building blocks of proteins, and assist in the production of protein from these amino acids, although ribosomes are also essential for this process. Enzymes also produce complex sugar or lipid molecules from smaller substrates; other enzymes degrade these macromolecules when necessary. A living cell would be nowhere without enzymes.

Proteins make up about 20% of the weight of a mammalian cell, or half of the dry weight of an *E. coli* cell. In the latter case, most of these proteins will be enzymes, although some proteins have different functions. For instance, ion channels (mentioned in the previous chapter) are made of proteins, which allow ion traffic through the membrane. This is not strictly an enzymatic activity, as there is no chemical reaction taking place. Other proteins bind to DNA to keep it in a particular conformation, again without actually catalyzing a chemical reaction. Nevertheless, the majority of bacterial proteins in a cell are enzymes, and all enzymes are proteins.

Many of the chemical reactions taking place in the cell can be performed in a reagent glass, too, but usually under very different conditions from those applying in a cell, such as high temperature, high pressure, or the presence of a catalyst or a solvent that would be toxic for a living being. The way enzymes catalyze chemical reactions in living organisms is very different from how we can chemically force the substrates to react together. It is therefore no surprise that natural enzymes are used in the laboratory, outside a cell, because they are so good at doing what they have to do under "normal" conditions. Enzymes are also used in industrial processes, and every household uses enzymes.

Consider a fat stain in a silk garment. Fat does not dissolve in water, but detergents will force some fat to let go of the silk because they contain surfactants. Surfactants that are typically found in a detergent are molecules with two sides:

one side that likes to be surrounded by water molecules and another that likes to be surrounded by lipids. Surfactant molecules would try to bind to the fat molecules in the stain, surrounding them with their "lipid side," while their "water side" faces water. By forming a complex of multiple detergent molecules surrounding a fat molecule, the latter could be released from the silk. This is a slow and inefficient process. Alternatively, fats can be chemically degraded by reactive solvents, which would be much faster but which might not do the silk much good. Enzymes, on the other hand, nibble away at fat, catalyzing chemical reactions that do not normally take place in water at laundry temperatures, and leave the silk alone. Enzymes degrade the fat into smaller, water-soluble molecules without harming the garment. The addition of enzymes that degrade fats (called *lipases*) and enzymes that degrade proteins (called *proteases*) to laundry detergents has greatly improved their effectiveness. Although the first enzymes to be used in this way were isolated from animal pancreas extracts (in the digestive track, the pancreas gland produces these enzymes to degrade proteins and lipids in the food), their isolation from bacteria enabled cheap production on an industrial scale. Present-day detergents nearly always contain microbial proteases, to remove fat and protein stains.

Enzymes are used in industrial processes with astonishing diversity. They frequently replaced toxic components that were in use before enzymes were discovered; enzymes are nontoxic and just as efficient. The production of textiles was dependent on alkaline or oxidizing agents, but these are more and more replaced by enzymes. Leather tanning needs fewer sulfides after the introduction of enzymes. Starch is produced with fewer acids, and paper can be reused thanks to enzymes. The addition of enzymes to animal feed can even reduce the amount of animal waste, as more of the feed becomes available to the animal. The fermentation and brewing of food or drinks depends on enzymes. They inactivate toxins and degrade plastic. Enzymes help in the production of cheese, as in the production or processing of soft drinks, chocolate, coffee, candies, or ice cream. A stonewashed jeans owes its appearance to cellulase, the enzyme that degrades some of the prestained cellulose in the cotton. (Cellulase was already mentioned in Chapter 8.) The enzyme "glucose isomerase" can transform the sugar glucose into fructose, which has a sweeter taste, and belongs to the most valued enzymes. It is derived from Gram-positive *Streptomyces* bacteria (of the phylum Actinobacteria) and is used to produce sweeter high- fructose syrup from less sweet corn syrup. Most of the enzymes with industrial applications are produced by bacteria, sometimes because they possess them naturally or else because geneticists have modified the bacteria so that they produce enzymes they would not normally make. For the production of a few enzymes, yeast is more suitable than bacteria. Yeast is a unicellular eukaryote, and some features of its protein synthesis are fundamentally different from that of bacteria, which can be important for the production of particular enzymes.

Every prokaryotic and eukaryotic cell alike uses enzymes for all its processes, including for repair of damage to its components, such as DNA. There are enzymes (called *DNAses*) that nibble away DNA to remove "loose ends" that require replacement, or they destroy DNA from attacking viruses. Other enzymes restore the bonds in DNA when a strand breaks. Enzymes can cut open a bond at a specific site, for whatever reason. The enzyme that produces DNA from its nucleotide building blocks is called *DNA polymerase* (in contrast to its name ending, this enzyme does not split but builds). Geneticists have used all these cutters, gluers, and knitters to manipulate DNA in the laboratory, as the specific reactions they carry out would be near to impossible to perform by chemical processes independent of enzymes. Once DNA is isolated from a cell, it can be cut and glued together in the desired combination, to be reintroduced in the same cell or added to a different cell. That way, geneticists can add properties to organisms that they did not yet possess, and DNA-manipulating enzymes are valuable tools in this process. One exception is the production of synthetic DNA, where novel DNA sequences are made from single nucleotides. This is typically done by chemical processes very different from how DNA polymerase does the job in the cell. However, once you have a piece of DNA and you want more of it, DNA polymerase is a far cheaper and simpler alternative to make copies than to boost up chemical production. DNA polymerase can only copy DNA from an existing strand, but it can't build DNA without having a template.

One type of DNA polymerase deserves particular attention. It is known by the name "*Taq* polymerase" or simply *"Taq." Taq* stands for *Thermus aquaticus*. This Gram-negative marine bacterium belongs to the Deinococcus-Thermus, a phylum we have not encountered up till now. It lives in hot springs and prefers temperatures between 50 and 80°C (120–175°F) for growth: it is a *thermophile* (heat lover). Its enzymes, including DNA polymerase, perform optimally at 70°C. Its discovery, plus a brilliant idea, produced a breakthrough in genetics, microbiology, molecular biology, biotechnology, forensics, medicine, taxonomy, paleobiology, and many other disciplines. Three letters sum it up: PCR.

The brilliant idea formed in 1985 in the head of Kary B. Mullis (United States), who realized that you could use DNA polymerase to multiply specific DNA fragments. Others had used DNA polymerase isolated from *E. coli* in the lab. The double helix of DNA can be split into its two strands, which in the cell is done by proteins but in the laboratory by simply heating the DNA. By giving it nucleotides, DNA polymerase uses one of these DNA strands as a template to produce the other. It does so from both strands, eventually producing two complete DNA copies where there had been one copy to start with. The protein keeps extending a DNA strand as long as it can follow a complementary strand as a template. However, DNA polymerase has to start somewhere, and you can make it start at a particular position by giving it a small piece of synthetic DNA, a "primer," to begin with. From that fixed position onwards, the enzyme will just keep going until it reaches the end of its template molecule.

Mullis realized that, if one separated the resulting DNA into the two strands and then gave the enzyme another fixed primer to run in the opposite direction, one could define both ends of the DNA to be produced by defining the position of the two primers used. The product would now be of a predicted length and defined exactly by the nature of the two primers chosen. Mullis further realized that the procedure could be repeated, whereby the formed DNA strands can serve as a template in the next step. If one were to repeat the reaction over and over again, more and more DNA of the desired length would be produced, or "amplified," in a chain reaction. One could produce multiple copies of a defined piece of DNA with exact beginning and end points. The idea of Polymeric Chain Reaction, *PCR* for short, was born.

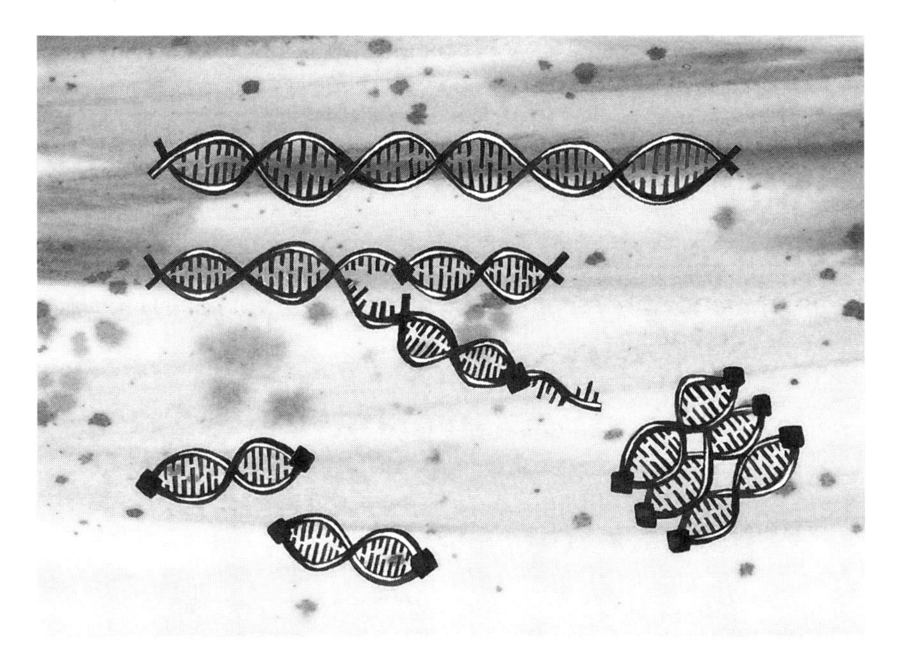

Mullis tried his method, and it worked, but it was tedious. You had to heat DNA to separate the two strands and cool it so that the primers could bind to the correct half molecules. Then you added *E. coli* DNA polymerase (the only DNA polymerase available in purified form, at that time) and kept the solution at 37°C, the optimal temperature for this enzyme. The next step, heating, separated the strands again, but it inactivated the *E. coli* enzyme, so new DNA polymerase had to be added to continue. Enter *Taq* polymerase. This enzyme not only survived the heating step but also performed its task at a temperature higher than *E. coli* polymerase, which was even better for the speed of the reaction. Soon, machines that changed the temperature of a reagent tube were developed, so one did not have to change the tube from one water bath to the next in order to vary the incubation temperature. The discovery of PCR revolutionized molecular biology and made DNA manipulation available to a much broader scientific audience than geneticists and molecular biologists, including microbiologists.

Mullis received a Nobel Prize for his discovery in 1993. Nowadays, PCR is routinely used in many, many applications. It can produce lots of DNA from samples containing only a single molecule, and it amplifies exactly that piece of DNA you want. PCR is now used to detect the presence of pathogens, by detecting a specific bit of their telltale DNA, in patients' samples, such as sputum, blood, urine, feces, or biopsies. It has speeded up medical diagnosis and improved the precision of the tests. PCR is also used to detect pathogens in food, so that products in the food chain can be checked at a higher speed and accuracy for the potential presence of disease causing bugs, thus improving food safety. By PCR amplification, the fragmented DNA of a Neanderthal fossil was amplified and sequenced, which revealed how similar its genome must have been to ours. PCR was used to demonstrate that people living in the Andes have suffered from tuberculosis for centuries, as mummies that were over 1000 years old had *Mycobacterium tuberculosis* DNA in their cells. DNA can be produced from a single hair or from a skin cell recovered from a crime scene, after which it can be matched to that of suspects (or to the police officer who took the sample, or the laboratory assistant who run the test, if insufficient care is taken to avoid such "contamination"). For microbiologists, it is important that PCR is used to amplify the DNA of bacteria that we cannot culture but still want to study. Their complete genome can be amplified and sequenced without a single cell dividing in the laboratory. The list of applications in which the PCR technology is essential is only restricted by the creativity of the scientists using it.

In bacterial taxonomy (explained in Chapter 2), the gene that codes for the RNA that is part of the ribosome (the amino acid knitter that produces protein) is particularly important. Ribosomes are made up of a number of different proteins and three pieces of RNA. The RNA pieces, and thus the DNA genes that produce these, differ significantly between eukaryotes and prokaryotes; within prokaryotes, there are again extensive differences between archaea and eubacteria. Lastly, within the Eubacteria or Archaea domains, isolates can be grouped into phyla, families, and genera, largely based on the slight variations encountered in their ribosomal RNA. The gene for one of these ribosomal RNAs can be amplified by PCR using universal primers that specifically fit on bacterial DNA (a different set of primers is needed for archaeal ribosomal RNA, which differs in sequence). After the correct fragment of DNA has been amplified, it is sequenced to determine to which phylum, family, or genus the organism from which the DNA was isolated belongs. In many cases, even the species can accurately be predicted from the ribosomal RNA gene. This is how we can state which species or genera are present in ecosystems without culturing the bacteria: we perform PCR amplification of isolated DNA and sequence the ribosomal RNA gene present.

How important enzymes are to manipulate DNA in a laboratory can hardly be exaggerated. PCR is only one of many techniques that have revolutionized life sciences, including genetics. If it was not for DNA-modifying enzymes, we would be unable to isolate genes or identify their function. In the next chapter, some of the basics of genetics will be explained. Biotechnology is the field where enzymes are needed, produced, modified, improved, and studied. Most biotechnological processes depend, directly or indirectly, on bacteria. If these bacteria were to go on strike, our industries would be in serious trouble. Luckily, this will not happen. As long as we feed them and care for them, bacteria will do their jobs without complaining. Praise to these silent workers of our society, and their enzymes that know their chemistry so well.

PCR in forensics

Forensic investigations can no longer be imagined without the use of PCR to produce a "DNA fingerprint" of human material obtained from a crime scene, be it from hair, skin cells, saliva, or other body fluids. A tiny bit of DNA is partly purified from any of these sources and then amplified with the help of primers that are specific to human DNA. These primers recognize highly repetitive sequences that our DNA contains at multiple locations. The distance between these repeat locations is different in every individual, which is what produces the individual DNA fingerprint, as the PCR products will be of different length depending on the distance of the primer's locations. The DNA fingerprint obtained is then compared with that of potential culprits. However, DNA evidence is not without its difficulties.

It is one thing to produce a DNA fingerprint, but another to match it with certainty to the correct individual. The statistics of a match being "significant" are often misinterpreted by judges and attorneys alike, and even when the DNA matches that of an individual with a high degree of certainty, this only implies the person was most probably present at the scene. It does not necessarily prove that he or she is guilty of the crime itself.

The generated DNA fingerprints do not provide any information on genes, although some information on the geographical origin of the individual, and their sex, can sometimes be deduced.

A recent DNA puzzle in Germany illustrates some of the pitfalls of DNA forensics. An identical DNA fingerprint was obtained from various crime scenes all over the country, multiple years apart, which varied from minor burglaries to six cases of homicide, including the killing of a police officer. The DNA belonged to a woman, which made it an even more curious case. How could this "phantom female killer" be guilty of over 40 minor and major crimes geographically and temporally so dispersed and be so reckless to leave DNA behind as if she did not care? The puzzle was eventually solved when it became clear that the DNA recovered was not from the culprit but was the result of contamination of the cotton swabs used to take the samples. The woman who had manually packed these swabs, that were not suitable for forensic investigations as they were not guaranteed DNA-free, had left her genetic fingerprint. The German police found out the hard way that saving money by buying cheap swabs is not a good idea.

11

Genetics and Genomics

When my friend Joe was a little boy, he loved to play with toy cars, especially Dinky toys. The more expensive ones had doors that could open, and tires that could come off the wheels. Taking all loose parts off was his favorite activity. His mother used to take him on shopping tours, which he hated, but she could bribe him with the promise of a new car. One time, when they were queuing to pay at the counter of a large store, a shiny new car burning in Joe's little hand, his mother remarked: "Shall I have it wrapped or are you going to destroy it right away?" He became a great geneticist working in microbiology.

Geneticists study the function of genes, and they frequently do so by destroying their gene of interest. So far, it has not actually been explained or defined what a gene is. Basically, a gene is a segment of DNA that bears the information to produce a particular product. In most cases the gene product is a protein, whereby an RNA molecule is produced as an intermediate. Sometimes, as in ribosomal RNA, the RNA is the final product. The information to make the product is stored in the form of the order, or *sequence*, by which the four building blocks of DNA are arranged. The sequence of the DNA that makes up a protein gene defines which amino acids are used, and in which order, to produce that protein. Likewise, the sequence of the DNA that produces an RNA product defines the order of the building blocks producing that RNA. A gene has a beginning and an end, both of which are recognized by the proteins that produce the product by "reading" the gene from the cell's DNA. A cell's DNA contains lots of genes, which are arranged like beads on a string. Some genes are separated by stretches of DNA that seem

Bacteria: The Benign, the Bad, and the Beautiful, First Edition. Trudy M. Wassenaar.

to do nothing other than just separate those genes. However, this appearance may be deceptive: the DNA separating one gene from the next may contain information that instructs the proteins of the cell when and how much product of each gene is needed. Bacteria and eukaryotes use different signals for this, and details of how DNA sequences form genes may differ; however, essentially every gene produces a product, and geneticists study both the genes and their products.

Consider a student, called Liz, studying a pathogenic bacterium. Liz has discovered it probably produces a pore-forming toxin. She mixed the bacteria with red blood cells, and noticed the red hemoglobin being rapidly released from the damaged cells, applying the hemolysis test as explained in Chapter 9. Liz could even see this happen when she mixed erythrocytes only the liquid media in which she had cultured the pathogen, without adding the bacteria themselves. She had centrifuged the liquid culture to separate the bacteria from the media, and adding this liquid to the erythrocytes had already caused hemolysis. However, such culture media would cause hemolysis only after the pathogen had grown in it. So there seemed to be a factor with hemolytic activity that the bacteria produce when they grow, and they shed this into their medium. Which genes would they need to produce this hemolytic activity? Liz has to destroy the right gene and see if hemolysis would be prevented that way.

Genes can be inactivated by UV irradiation. These rays damage DNA, which is why overexposure to sunlight can cause skin cancer in the long term: the DNA in some skin cells can become irreparably damaged and eventually this can turn them into cancer cells. Bacterial DNA is also sensitive to UV. By exposing a bacterial culture to a short burst of UV light, you can damage on average only one gene in each cell. Now you let the bacteria grow on agar plates that contain blood. Only those bacteria whose DNA is not lethally damaged can grow, and they will produce colonies (heaps of bacterial cells) on the agar plates. As long as they can produce the hemolysin, each colony is surrounded by a colorless ring in the agar where red blood cells were destroyed and the bacteria could digest their content, including the red hemoglobin. But a few colonies may not produce such a ring: they have lost their ability to produce the hemolysin. You have made hemolysin-negative *mutants*. This is how geneticists produced mutants in the past, but it is far from easy to actually work out which gene it was that had been UV damaged.

Nowadays, there are better tools than UV to produce mutants. Geneticists frequently use naturally occurring DNA fragments that spontaneously insert themselves into bacterial DNA. These can be considered "parasitic DNA." When they insert themselves in-between two genes, nothing much might happen, but when they insert themselves within a gene, that gene can no longer work properly. These "DNA bullets," which are called *transposons*, are a favorite tool to destroy genes, especially when they carry an antibiotic resistance gene, which they frequently do. Imagine this experiment: the student Liz lets a transposon "attack" the pathogen of interest, and then grows the resulting bacteria on a blood agar plate containing the antibiotic that kills the pathogen, unless it has taken up the transposon, which bears a gene that renders resistance to the antibiotic. In that case, the recipient bacterial cell is not affected by the antibiotic and it will grow to produce a colony. Now she looks for colonies without the white zone, and, bingo, the transposon is likely to be present in a gene needed to produce the hemolytic activity. Look for the transposon in the genome (for instance, using PCR) and you have found the gene where it has been inserted.

After a few more controls, the industrious student has identified the gene whose product is the hemolysin. Now Liz wants to know if that toxin is actually responsible for the disease this pathogen causes: she wants to know if this gene is a *virulence gene*. The best way to do this is to show that the hemolysin-negative mutant can no longer cause disease. It would be unethical to let the mutant infect humans (although this depends on the severity of the disease, and human volunteers are used occasionally in such studies) but if a laboratory animal can get sick from the bacteria in a manner similar to how humans do, these provide a suitable animal model. Liz can test if the mutants have lost their ability to cause disease. In that case, she has provided evidence that the hemolysin gene is a virulence gene, although, again, more checks and controls are needed before she can convince her supervisor of that conclusion.

Apart from destroying genes, geneticists also add genes to an organism. To be able to do so, they use all those DNA manipulating enzymes that bacteria produce naturally, as was mentioned in the previous chapter, and which can conveniently be bought in pure form. Liz proposes to her supervisor to add the hemolysin gene to another bacterial species that is not normally hemolytic, to see if the addition of this gene can transfer that ability. It will not necessarily do so. You might have removed the ability to produce hemolysin by inactivating one single gene, but at the same time the cell might need several genes in order to produce active hemolysin. Besides the hemolysin gene itself, maybe the pathogen needs a gene coding for a chaperone to fold the protein in the right conformation, or a gene to specifically splice the protein to make it active, or another gene to transport it across the membrane. If that were the case, you could only transfer the complete activity, or the hemolytic *phenotype*, by transferring all the required genes to another species that naturally lacks these genes. Liz wants to become a molecular biologist and during her study she learns how to produce recombinant DNA. She performs all experiments to prove the Koch's postulates (mentioned in Chapter 1) that are the basis of research into pathogens. The transfer of her virulence gene to a foreign genetic background is one of her first recombinant DNA experiments in what promises to be a long, successful career.

One more piece of information is needed to produce a good scientific publication: do the bacteria produce the hemolysin all the time or under specific conditions only? Liz wants to vary the growth conditions of the pathogen, but realizes that some of the tested conditions would be detrimental to the erythrocytes that she needs to test hemolytic activity. She decides to replace the hemolysin gene with a *reporter* gene, by means of recombinant DNA techniques. She inserts this reporter gene that codes for the production of a blue pigment, exactly in the position where the hemolysin gene normally resides. All surrounding DNA is left untouched, including the genetic signals that tell the cell when to produce a gene product and when not to. Now she varies the growth conditions of her mutant and marks when the pigment is produced, which colors the bacteria dark blue. These are the conditions that are likely stimulating hemolysin production in the pathogen. Nowadays, a wide collection of reporter genes are available that produce a colorful palette; they are typically used to identify where and when a gene is expressed.

Recombinant DNA, that is, DNA with genes artificially manipulated by human activity, was very controversial when the techniques were developed. Is it OK to fiddle with DNA like this? Are there dangers involved, is it ethical, are we not descending a slippery slope? What can a rogue scientist do with the technique? These questions were asked when the first DNA manipulation enzymes were discovered, and the very first experiments were carried out to produce mutants, in the early 1970s. Scientists were worried that dangerous bacteria could be produced, deliberately or by accident, that novel diseases might break out, or that bacteria would escape from the laboratory and start to grow uncontrollably. In 1974, the scientific community collectively decided to pause all research involving recombinant DNA and to establish an advisory committee to assess the risks, while in the general public an anti-science attitude germinated. Graffiti on a wall in the Dutch university town of Leiden summed this up. It read "stop DNA."

The storm calmed only slowly. Scientists agreed to restrict their experiments to bacteria that were crippled for a start, so that in case they escaped none would survive outside the laboratory. Strict regulatory guidelines were defined in 1976, which were slowly relaxed as it became evident that the presumed risks were not as serious as feared. By 1979, scientists were permitted to work with DNA from cancer viruses, which had been previously excluded. Recombinant DNA techniques became accepted in the 1980s, and have stayed with us since.

Fortunately, there are internationally accepted safety rules about working with recombinant DNA that are mostly dictated by the pathogenic potential of the organism whose DNA is modified. Strict safety rules apply to pathogens causing dangerous diseases, such as *Bacillus anthracis* or *Yersinia pestis*. These are handled in a Biosafety Level Three laboratory. The highest level is four, where personnel wear protective clothing, gloves, and masks; entrances are double-door airlocks; and the laboratory air is under-pressurized. All air, water, and waste leaving the laboratory is decontaminated before being released, and personnel have to shower and change upon entry and departure to ensure the safety of both the workers and the environment. Biosafety Level Four is reserved for deadly diseases for which there is no treatment or vaccine, and which can be transmitted by aerosols, for instance, Ebola or smallpox (both are caused by viruses). Less rigid safety rules apply to organisms that can cause less serious diseases. While working with recombinant organisms that do not cause disease, the general safety rules of any bacteriological laboratory apply: no foods or drinks, hand washing after handling biological material, and compulsory use of laboratory coats. Biological material is usually sterilized before disposal.

The term *genome* was frequently used so far, but what is the difference between a genome and a gene? A gene is a DNA segment containing all genetic information for one protein (or RNA molecule), whereas all DNA of a given cell comprises its genome. The genome describes all DNA in a given cell, in which all genes will be found. As genetic insights progressed, scientists became interested in the genome as much as in individual genes, to get the overall genetic picture of a cell. Several techniques exist to determine the sequence of the four nucleotides from which DNA is built, and by knowing this sequence one would, in principle, know what the DNA is coding for. In practice, we do not always know even if we can read the sequence, like reading a foreign language without understanding the words. With technology advancing, DNA sequencing became faster and cheaper, so that in 1995 the complete genome sequence of a living organism was published: that of *Haemophilus influenzae*. These Gram-negative bacteria of the phylum Proteobacteria were originally thought to cause influenza. They do not, but instead are normal inhabitants of the nose or throat, and occasionally cause ear infections, sinusitis, or airway infections. This organism was chosen to be completely sequenced because of its relatively small genome, consisting of 1800 kb only (a nucleotide is a base and a kilobase, or kb, is 1000 bases). It took the research group a year to sequence all this DNA, back then. A genome ten times that size can now be sequenced by automated machines within a day. As a result of this major technological advance, we have recently hailed the thousandth bacterial genome to have been completely sequenced.

With the first genome sequence released, the field of *genomics* was born. It is a mixture of laboratory bench work and computer analysis, as the long DNA sequences can only be analyzed with the help of specialized computer programs. Bioinformatics is another new star to the scientific firmament: programmers, mathematicians, and informaticians have specialized in analyzing biological sequences. They can even do experiments, where supercomputers have replaced the agar plates and test tubes, and perform complex calculations to identify genes, predict their function, or assess their conservation between organisms. Genomics has become a fascinating specialization within microbiology, just like it is producing breakthroughs in classical biology, as the genomes of more and more plants and animals are sequenced.

Where will this lead us? We can sequence the genome of bacteria within a day, and this applies to the most dangerous pathogens, too. DNA can be produced synthetically, and multiplied by PCR. That combination can potentially lead to dangerous situations. There have been discussions whether the genome sequence of highly pathogenic organisms should remain secret and not be made public, so that this information cannot be misused. Many researchers believe this would not be a good idea. Openness allows more people to gain experience. Knowing that *B. anthracis* can be isolated from soil without too much trouble, and its genome sequence could easily be determined by a bioterrorist with malicious intents, it is better that many scientists have access to this information, than to keep it restricted. The latter would only help rogue scientists keep their scientific advances hidden and gain a lead. An example of the use of genomics to solve such crime is presented in Chapter 19.

The future of genetics and genomics will have a lot in store. It is expected that within a few decades, so many bacterial genomes will have been sequenced that computers will no longer be able to store, let alone analyze, that vast amount of data. New approaches are being discussed that still sound like science fiction, but may become true earlier than expected. Maybe, in a not so distant future, your doctor no longer sends a sputum sample to the laboratory to have bacteria cultured, but sticks the swab into a machine, to have a complete genome of the pathogen a few hours later. Let us hope the doctor knows what to do next, and that there still will be effective drugs. The next chapter shows that the latter cannot be guaranteed.

12

Antibiotics and Resistance

The discovery of antibiotic substances has produced two Nobel Prizes (see Chapter 7), and, at that time, resulted in a confident attitude. Some optimists predicted that soon all bacterial infections would be treatable, and a pessimistic microbiologist could think that (medical) microbial research would eventually die out from lack of interest. This futurism was dampened when the first bacteria were discovered that had become resistant, and in the following decades the fear that antibiotics would one day become useless proved to have some ground. Before discussing resistance, we first take a closer look at what antibiotics are, and what they do.

Antibiotics are compounds that kill bacteria, or inhibit their growth, which can be used to treat bacterial infections since they do not harm the diseased host. A lot of compounds are toxic to bacteria, but those that are also toxic to animals or humans are not suitable to treat a patient. Antibiotics can act specifically against bacteria because the details of cellular processes taking place in bacteria differ from those happening in eukaryotic, and thus in human and animal, cells. As long as antibiotics attack bacterial processes that our own cells carry out in a different way, the drugs will have limited effects on our own body, and the gain of combating infection outweighs by far the minor risk of mild side effects that may occur.

As was discussed in Chapter 9, bacterial toxins frequently attack the membrane of their target cell, or else interfere with protein synthesis, and these are the targets for many antibiotics as well, this time upsetting bacterial cell membrane or protein synthesis. Obviously, inhibitors of cell membrane or protein synthesis would have

---

Bacteria: The Benign, the Bad, and the Beautiful, First Edition. Trudy M. Wassenaar.

no effect on viruses, which produce neither, so antibiotics do not work for viral diseases. Nevertheless, there are drugs that can inhibit viral reproduction, and these antivirals are members of the broader group of drugs called *antimicrobials* but they are not antibiotics, a term restricted to compounds active against bacteria.

The cell membrane of Gram-positive bacteria is surrounded by a structure called a *cell wall* that is composed mainly of a thick layer of peptidoglycan polymers (long chains of sugar molecules) that are cross-linked into a net-like structure. Gram-positives are the only organisms to produce such a cell wall, making it a suitable target for an antibiotic. Indeed, penicillin, the first antibiotic that was discovered and used in medicine, specifically inhibits the interconnection of peptidoglycan polymers, by inhibiting the specific enzyme that makes the connections. It means that penicillin does not affect an existing cell wall, but in its presence bacteria can no longer produce new, stable cell walls, so they can no longer grow. Penicillin does not kill as much as it prevents bacterial growth. This is sufficient to treat an acute infection, since the defenses of our own body can soon take a lead when the pathogen is no longer multiplying.

It is even known how exactly penicillin inhibits the cross-linking of peptidoglycan. Its chemical structure contains a "lactam" group (penicillin belongs to the antibiotic family of beta-lactams). This allows it to bind to the target enzyme after which it gets stuck like a wrong key in a lock: it will not let go of the enzyme. Every enzyme molecule to which penicillin is bound can no longer do its proper job of interlinking peptidoglycan polymers. Penicillin even works against some Gram-negative bacteria, since these also produce peptidoglycan, although they do not use it for a cell wall. The details differ, but the result is the same: penicillin stops the bacteria from growing.

How does the drug that we take when we are sick know which bacteria to attack and which to leave alone? It does not, and all bacteria in your body that are exposed to the drug, and not resistant or insensitive to it, will either stop growing or be killed, when they receive a high enough dose. Part of the gut's microflora will suffer from an antibiotic dose, and the relative numbers of different bacterial species will change, with diarrhea as a possible side effect. After the drug is no longer taken, the diarrhea usually disappears. Nevertheless, a changed distribution of bacterial species in the gut can be detected long after the antibiotic has left the body. It has been realized only recently that antibiotic use has long-term effects, whose implications we do not fully understand. A more serious side effect results when a patient is allergic to the drug, which is not uncommon in case of penicillin. In that case, it should be quickly replaced by an alternative antibiotic, as allergic reactions can at best be a nuisance, but at worse develop into a life-threatening condition.

Penicillin is not a human invention. It is produced by molds (fungi) and certain types of bacteria, which use it to inhibit bacteria growing on the nutrients they would like to reserve for themselves. Those competing bacteria, on the other hand, have learned a few tricks to combat their hostile cohabitants. Some bacteria produce an enzyme that breaks down penicillin, before it can harm them. This enzyme, penicillinase (also called betalactamase) is naturally made by a number of bacteria that are genuinely insensitive to penicillin. *Enterococcus* bacteria (these are members of the phylum Firmicutes), which naturally live in the intestine, are an example of Gram-positive bacteria that happily grow in the presence of penicillin.

Imagine what would happen if *Staphylococcus aureus*, a Firmicute bacterium normally living in the nose or throat, but able to cause nasty infections when introduced into a wound, would take up the penicillinase gene of *Enterococcus*. Now "Staph," as this organism is known for short, can no longer be treated with penicillin. A scenario like this happened after the introduction of penicillin. For Staph, it was easy to accept a gene that helped it escape the toxic effect of the drug, although we do not know for sure which bacteria donated the gene. Many antibiotic resistance genes reside on transposons (pieces of "mobile" DNA that can insert themselves into new genomes, introduced in the previous chapter), and geneticists have discovered how useful these are to transfer resistance from one cell to the next. Bacteria transfer transposons naturally, especially when they help them survive an otherwise toxic dose, by receiving a resistance gene or two. Penicillinase has been described as the enzyme that proved the most costly to society. Its transfer to novel species has resulted in many resistant pathogens that are now much harder to treat, causing innumerable suffering and frequent deaths.

Transposons, or other, likewise mobile DNA fragments, frequently carry more than one antibiotic resistance gene. In Chapter 5, it was explained how multiple genes can be found located together, if their combined transfer has evolutionary benefit, with the example of genes producing a TTSS. Antibiotic resistance genes can also clot together, and multiple genes may be present on a single transposon. Some strains of *Salmonella enterica*, for instance, contain a transposon in their DNA that carries five different genes that produce resistance to five antibiotics, each of which works differently. Staph has also learned this trick. The dreaded MRSA, short for methicillin-resistant *Staphylococcus aureus* is resistant to a number of frequently used antibiotics, not only methicillin (a penicillin-type antibiotic), which is why the MR in MRSA is sometimes read as "multiresistant." MRSA can still be treated with vancomycin, but some strains of the organism have also learned to overcome this antibiotic.

It is easy to imagine that resistant bacteria have an advantage in an environment where they frequently encounter antibiotics. Indeed, multiresistant bacteria are most often found in hospitals, where sick and weakened patients, whose immune systems are vulnerable, are easy victims, and antibiotics are frequently used. This is the tragedy behind hospital-acquired infections. Sloppy doctors or bad hygiene can be prevented, but the unfortunate combination of vulnerable patients, resistant bacteria, and abundant (but necessary) use of antibiotics will always pose a risk of hospital-acquired, resistant infections that are hard to treat.

An alternative mechanism by which bacteria can become resistant is also worth mentioning. Not only the uptake of a novel gene but also a change in an existing gene can result in resistance. For this, we will turn to the macrolide family of antibiotics, which inhibit bacterial protein synthesis by blocking the ribosomes (see Chapter 9 for details on protein synthesis and the role of ribosomes therein). Some bacteria can escape this antibiotic through a mutation in one of the genes that produce the ribosome. A change in a single DNA base, resulting from a mistake made by DNA polymerase, produces a functional ribosome that is now insensitive to macrolides, because the change prevents binding of the macrolide to the ribosome. Such point mutations are a less frequent mechanism resulting in resistance than gene uptake is, but it illustrates how powerful mutations can be. *Helicobacter pylori* can cause gastric ulcers in a minority of persons who carry these Gram-negative Proteobacteria in their stomach. The treatment with erythromycin (a macrolide) soon became ineffective, as the point mutation resulting in resistance occurs at a frequency high enough to produce at least one resistant cell in a typical colonized stomach. One resistant cell that survives is enough, as one bacterium can produce a population of millions within days.

Bacteria also have a further way to overcome the toxic effect of antibiotics: they can simply remove the toxin from their cells. By active transport through the membranes, toxic compounds are removed as part of the day-to-day homeostasis that cells perform, and by tuning up the genes coding for the pumps involved, bacteria can become less vulnerable to a wide variety of antibiotics. Such pumps can frequently dispose of various toxic compounds, so that producing a bit more of the pump protein makes the bacteria less susceptible to various antibiotics simultaneously.

It is one thing to be resistant to antibiotics, but another to be dependent on them. A most unusual discovery was made in a hospital in Sheffield (United Kingdom) in 1987. A road traffic accident victim was treated for various fractions at the intensive care unit, and for the next few weeks suffered from recurrent chest infections that were treated with various antibiotic courses. A sputum sample taken 37 days after admission was found to contain *Neisseria pharyngis* bacteria (a normal Gram-negative inhabitant of the throat) that would only grow when a mix of antibiotics was added to their media. Without antibiotics, they would not multiply: the bacteria apparently were dependent on the drugs, although these were of types different from the antibiotics the patient had received. It remains a mystery how these bacteria had become dependent on antibiotics for growth, and the observation is very uncommon, as few such bacteria have been discovered since. In the hypothetical event that these would cause an infection, the unusual treatment would be to take the patient off antibiotics.

The war against pathogenic bacteria with antibiotics as a weapon has become a battle between chemists and microorganisms. Chemists try to chemically change existing drugs so that they remain active even against bacteria that have developed resistance against the original compound. Bacteria explore novel ways to become resistant again in order to survive the toxicity of the antibiotics they encounter. It seems as if, sooner or later, pathogens will learn to cope with all antibiotics, as they have learned to cope with so many hostile challenges over evolutionary times.

Bacteria that have managed to become resistant to many clinically relevant antibiotics are described in the press as "superbugs." Superbug *E. coli* bacteria that are resistant to carbapenems, a class of very potent antibiotics that are used as a last resort, have emerged in hospitals in the United Kingdom, possibly after being imported from Asia, where over-the-counter sales and off-label use of antibiotics has resulted in serious resistance problems. Superbug *Enterococcus faecalis* that is resistant against vancomycin, another last-resort drug, is as much feared as the MRSA previously mentioned, and causes ongoing problems in hospitals around the world. Superbug *Mycobacterium tuberculosis* that is essentially non-treatable, as it is resistant to all relevant antibiotics, was detected for the first time in 2006. Patients infected with these bacteria can basically no longer be treated. That observation leads the way to what can be expected in the future if novel drugs are not quickly enough discovered, and antibiotics are not used with more care.

We can do our little bit to keep up with the bacteria. Antibiotics should not be used to treat viral infections, or to relieve mild symptoms when there is no real risk of things getting out of control. Antibiotics should not be taken longer than needed, and never at a lower than the required dose, as that gives the bacteria an even greater advantage to become resistant. If an antibiotic treatment doesn't work, the dosis should not be increased, but instead the drug should be replaced with another antibiotic, preferably one that functions in a different way. The use of antibacterial household products should be discouraged.

Maybe we should put a little more trust in the defenses of our own body. In historical times, bacterial infections were big killers, and in some countries they still are, but the introduction of antibiotics was only one reason why our chances are so much better now: vaccination, improved hygiene, and improved nutrition all have drastically changed the odds. Bacterial infections can still be lethal, and dangerous, but the vast majority of infections you will encounter in your life will be dealt with by your own body, without the need for antibiotic medication.

It is a tightrope walk. Some people decry all antibiotics as "poison" and will refuse their use categorically. That can be a dangerous strategy, and they have to accept the risk of suffering from long-term damage, or even dying from an infection that could have been easily treated. Other people insist on an antibiotic as a "magic pill" to comfort their symptoms, even if these are only mild. That can also be a dangerous strategy, as collectively it builds up resistance that one day can kill patients who rely on the drug. Doctors and patients alike can contribute to keep the balance that is needed to count on antibiotics for times to come

13

Marine Microbiology

In the first chapter, an extraterrestrial view was presented of the blue planet being the home of bacteria. The planet appears blue from outer space because of its vast amounts of water, and now it is time to take a look at life in our oceans: we will turn to marine microbiology. Bacteria living in the oceans are as abundant as those living on land. They can be swimming free; live on solid surfaces, animals, or water plants; or exist inside animals or their cells. The biology of the deep seas is far less well studied than that of land, and likewise, we know even less of marine microbiology than we know of terrestrial microbes. The snippets of information we have, though, paint a wonderful and colorful picture.

A short exploration of the conditions of this wet environment that covers nearly three-quarters of our planet will set the scene. The water of oceans and seas is salty and rich in minerals; these mostly originate from the ocean floor from which they are being released as a result of tectonic activity. The variation in salt content is generally small, although at the surface it can be diluted by excessive rain fall and by the influx of rivers in coastal areas, or it can be concentrated by ice formation (in the Arctic) or evaporation (in the tropics). Especially in enclosed waters such as the Mediterranean and the Red Sea, evaporation increases surface salt concentrations considerably, with the Dead Sea being exceptionally salty. The salt concentration also varies with depth and temperature, whereby cooler deep water (and cooler polar water) is usually less salty. The deep Atlantic, though, is relatively salty, even at depth. Light is available in the top 200 m at the most, deeper waters are pitch dark. Seawater contains dissolved oxygen, with the highest concentration in

Bacteria: The Benign, the Bad, and the Beautiful, First Edition. Trudy M. Wassenaar.

the top 20 m, decreasing gradually to an oxygen-depleted zone at a depth of around 1000 m. Oxygen then increases again with depth to reach concentrations in deep waters similar to those found at sea level. It means that the deep sea (75% of all sea water is below 1000 m) is not anoxic, once you have passed the oxygen-depleted zone. A further variable is the hydrostatic pressure, which increases by 1 atm for every 10 m of depth, and can reach 1000 atm or more for the deepest places in the ocean. The average depth of oceans is 3.8 km, but the ocean floor exhibits mountain ridges as well as deep trenches. The Marianas Trench in the Pacific is the deepest, reaching 11 km below surface level. Temperature is also variable, and decreases with depth. The temperature of the oceans is mostly between 2 and 10°C and relatively constant at a given depth. This means that many bacteria living in the major habitat of our planet, the oceans, normally grow at temperatures below 10°C. The temperature in the top 200 m strongly varies with location. Surface water in the tropics can reach 30°C, whereas in the arctic zones it can freeze into sea ice, which means it must be colder than −1.9°C.

The water of our oceans circulates both horizontally and vertically with large-scale ocean currents. Nevertheless, one ocean is not like another, and location matters, not only geographically but also in terms of depth. The bacteria that live in the ocean also vary according to location and depth. Those bacteria living at high pressure in deep water can only be cultured in pressure chambers. Most organic matter, however, is found in the form of plankton in a thin layer at the surface. A constant supply of organic particles, derived from dead cells and debris, falls down from this fertile zone, which is termed *marine snow* because the particles aggregate to resemble snowflakes. These flakes are home to large numbers of bacteria.

As was explained in Chapter 6, all life depends on an external energy source. Marine bacteria depend either on light or on chemical energy, but the lack of light in deeper waters exclude photosynthetic bacteria in most of the water. Nevertheless, we will first consider life that depends on light.

Marine bacteria are often smaller than their terrestrial counterparts, but nevertheless form a considerable proportion of the plankton that lives in the top layer of all oceans. Plankton is a collective name for photosynthetic bacteria and archaea, together with eukaryotic algae and protists feeding on these (flagellates, dinoflagellates, diatoms, and the like). Larvae of marine animals are also common members of the plankton ecosystem. Photosynthetic bacteria (mostly Cyanobacteria), archaea, and algae are the "lungs" of the oceans, as they convert carbon dioxide into biomatter with the help of light, thereby producing oxygen. The total amount of oxygen produced by marine organisms exceeds that produced by land plants and trees. Also, counterintuitively, most oxygen is produced in moderate, not in warmer marine zones. This is because cold water contains more nutrients, and deep, cold, nutrient-rich water rises to the surface (where light is plentiful) in vast quantities in the moderate temperature zones. The relatively low amount of plankton in tropical zones is the reason why those waters are so clear. The deep blue, transparent sea you may dream of for your holiday is in fact a microbiological "starvation zone."

The amount of plankton is limited by the availability of nutrients, the amount of sunlight, the temperature of the water, and by the organisms that prey on the planktonic microorganisms. Sometimes things run out of control and plankton quantities explode. This is called *plankton bloom*, where huge mats of often colorful plankton cover large areas of water. Plankton bloom can occur both in seas and in freshwater lakes, and extensive blooms are even visible from outer space. They can be quick events, lasting for days only, or take weeks to disappear. Some blooms are seasonal and predictable, others occur unexpectedly. Many factors must come together before the population of planktonic organisms can explode, but their appearance and disappearance are both stressful events to an ecosystem, as local equilibriums are dramatically disturbed. Algal bloom in lakes and coastal waters can be problematic because many Cyanobacteria, for instance, *Microcystis aeruginosa*, produce toxic substances (this is only one of at least 40 genera that are known to do so), and during bloom these reach toxic concentrations. This not only prohibits swimming, but can kill birds, fish, and mammals alike. Moreover, the toxins accumulate in shellfish, so that their consumption becomes dangerous. By the time the plankton dies, its decomposition produces a foul smell, but more seriously, the water can locally become anoxic because the bacteria living off this organic feast are using massive amounts of oxygen, with devastating effects on the ecosystem. Tourism can be greatly impaired and the fishing industry suffers when the plankton goes into overdrive. Some plankton blooms are the result of human activity, mostly by agricultural practices, where overused fertilizers are washed away into the rivers and eventually reach the sea. This results in the unintentional increase in the amount of available nutrients. However, not all plankton blooms are the result of human activity, as some are natural phenomena.

Despite the negative effects of occasional bloom formation, photosynthetic plankton is the basis of the marine food chain. We will come to photosynthesis in marine bacteria, but since most people are familiar with the process in plants, this will first be introduced. Land and water plants alike use chlorophyll to perform photosynthesis, which resides in their chloroplasts (see Chapter 6 on the origin of these organelles inside plant and algal cells). Chlorophyll is a pigment that contains a chemical *heme* structure similar to that of hemoglobin in our blood, but is green instead of red. Chlorophyll has a remarkable ability: it captures the energy of visible light and converts it into chemical energy. With the help of chlorophyll, the chloroplasts of plants can produce sugar from water and dissolved carbon dioxide, a process for which a lot of energy is needed. This energy demand is easy to understand by reversing the chemical reaction: burning sugar produces carbon dioxide and energy, which is what our own cells do, just like cells of other animals (and even plants). When burning sugar releases energy, producing it must cost energy, which plants get from sunlight. The conversion of carbon dioxide into sugars, and subsequently into biomatter, is called *carbon fixation*.

The by-product of photosynthesis that turns carbon dioxide into sugar is oxygen, and this is produced as follows. Plant chloroplasts, as a first step, strip water molecules of their hydrogen (the "H" in H_2O) and catch energy from light with their chlorophyll. The energy is temporarily stored in the form of ATP. This cellular energy storage molecule serves as a rechargeable battery. The valuable hydrogen, needed to make sugar from carbon dioxide, is kept as hydrogen ions, and is bound onto a carrier for later use. To put this hydrogen aside like this again needs light, by the action of chlorophyll. The splitting of water also releases two electrons, which enter a "bandwagon" of reactions to eventually be dropped at two oxygen atoms that were once part of water molecules; eventually this oxygen is released as dioxygen gas, O_2.

All components come together in a second series of chemical reactions to produce the much-wanted sugar. This second reaction chain can either happen simultaneously with the first part, or even after it gets dark, because all necessary energy is now stored. During the final steps, the carbon from carbon dioxide will combine with the stored hydrogen and grow into a sugar, while the stored ATP provides the energy for this feat (Chapter 14 deals with some of the details of this part). This is how the chloroplasts in the cells of plants and algae perform photosynthesis, on land as well as in the sea. It sounds complex, and it is. But, as we have seen in Chapter 6, plants have not learned this difficult task themselves: they imported it with the endosymbiont Cyanobacteria that eventually evolved into chloroplasts. The process of producing sugar from carbon dioxide and water with the use of light as an energy source is originally a bacterial invention.

Not only plants perform photosynthesis. Independently living Cyanobacteria (previously called "blue-green algae") also still use this strategy, and do so in large quantities; *Prochlorococcus* and *Synechococcus* species are abundant in the ocean. These and other marine Cyanobacteria perform photosynthesis every day, and produce lots of oxygen as they grow, but they are not the only organisms that use sunlight as an energy source to produce sugar.

The discoveries of photosynthesis and chlorophyll in plants were made by successive investigators and the details were long known before it was discovered that bacteria other than Cyanobacteria can convert light into energy as well. A group within the phylum Proteobacteria, called *purple bacteria*, were found to be able to perform photosynthesis, for instance, bacteria of the genus *Rhodospirillum* and others that have names starting with "Rhodo," meaning rose-(colored). They do not depend on chlorophyll but use an alternative pigment called *bacteriochlorophyll* for light harvesting, which is related to, but different from, the chlorophyll of plants and Cyanobacteria. The pigments they need to perform all necessary reactions give these bacteria their bright colors. Apart from Proteobacteria, members of the phyla Chlorobi and Chloroflexi also use bacteriochlorophyll to produce sugars with the use of light. Not all of these photosynthesizers produce oxygen as a by-product, and most do not use dissolved carbon dioxide as a carbon source, but instead favor simple organic molecules. In that case, the organism is called a *phototroph*, meaning its energy source is light, but it does not perform carbon fixation, so they are not autotrophs (self-feeding, see Chapter 6). This, some biologists would argue, disqualifies them as being truly photosynthetic, although they synthesize sugars with the use of light. Instead of stripping hydrogen from water molecules (H_2O), some bacteria may use hydrogen disulfide (H_2S). But they all use light as their energy source with bacteriochlorophyll as the light-harvesting pigment. Their key issue is to produce sugars, which they need for growth.

There is an even simpler way to extract energy from light, and that is possibly the oldest biological "invention," although it was discovered last: the use of bacteriorhodopsin. This pigment is purple, and it chemically resembles the light-sensitive rhodopsin in our eyes. Instead of vision, bacteriorhodopsin produces energy for the bacteria and archaea that produce it, and this energy is used for the production of sugar from a carbon source. Bacteria and archaea that use bacteriorhodopsin do not need the electron-transport bandwagon that is needed when chlorophyll or bacteriochlorophyll capture light energy. In its simplest form, bacteriorhodopsin catches light to produce ATP, which provides the cell with energy for sugar production.

Archaea are the main prokaryotes living in the oceans below 100 m, and as such, archaea are major inhabitants of oceans. They are estimated to make up one-third of all prokaryotic life in the global ocean. Archaea play a major part in the biological recycling of nitrogen, an important nutrient that is recycled between the atmosphere and the biosphere, as seen in Chapter 18. There is some indirect evidence that marine archaea may also fix inorganic carbon. Those archaea that live in the deep sea must use alternative energy sources to photosynthesis, as they live in absolute darkness, but since these organisms cannot be cultured, not much is currently known about their metabolism.

We will continue with light, but now produced instead of used by marine bacteria. Light-emitting bacteria are widespread in the dark ocean, where they receive far more attraction than on land. Bioluminescence, as the production of light by living organisms is called, can literally be seen in a dark night as a soft glow in the water, when enough bacteria simultaneously do it. It is practiced by bacteria having a symbiotic, parasitic or free-living lifestyle. Many bioluminescent bacteria live in symbiosis with fish, where they can be found in specialized organs, and these light bulbs on the animal's body attract the prey that the fish are after. The bacteria must be taken good care of by their fish host, as luminescence can use up as much as 20% of bacterial energy consumption, and the fish must feed these endosymbionts. The bacteria shine by the use of an enzyme called luciferase, which emits light while performing an energy-consuming oxidation reaction. *Vibrio fischeri*, a distant cousin of *Vibrio cholerae*, lives as a symbiont in Pinecone fish (*Monocentridae*) and squids (*Sepiola*) and is a well-studied lighting subject.

The most famous luciferase is green fluorescent protein, or GFP, which jellyfish (*Aequorea victoria*) produce without the help of bacteria. The protein was discovered by Martin Chalfie, and resulted in a shared Nobel Prize in Chemistry, in 2008. He was the first person who was able to introduce the gene for GFP into *E. coli*, producing shiny green bacteria under a UV lamp. In contrast to the common belief that UV light does not penetrate water, there is enough UV light in deep water to excite the protein and let marine organisms shine. Nowadays, GFP is widely used to investigate where and when genes are used within an organism; it is used as a marker gene, like pigment genes described in Chapter 11. You can tag a protein of interest with GFP by genetic procedures, and then wait to see which parts of the organism light up. The light will tell you where and when the protein is being produced. Scientists thus produced, as a proof-of-principle experiment, a mouse whose body illuminated, but it was not well received by the public. Ghostly light-emitting mice are not appealing to many people. Nowadays, various fluorescent proteins are used as marker genes, and under a microscope one can observe genes in action, by the appearance of fluorescent light signals even inside a cell.

Returning to the sea, we will have a look at the marine equivalent of a terrestrial bacteria-rich compost heap or waste dump: whale carcasses. Whales are the biggest mammals on earth and a dead whale will sink to the bottom of the ocean, where it takes years to decompose, as there are few large scavengers at this depth. A whale carcass is a nirvana for bacteria, and for the few clams, worms, and shrimps that are specialized to live at this depth. The animals may nibble at the flesh, but only bacteria decompose both the flesh and the skeleton, thereby producing hydrogen sulfide (H_2S, which above water strongly smells of rotten eggs), which again is food for other bacteria, as we have seen above. A dead whale is an ecosystem full of life. By the time it is reduced to a skeleton, the outline of the disappeared massive body is still visible on the ocean's floor as a shadow of bacterial growth.

Whale carcasses may seem a strange biotope, but cold seeps, hydrothermal vents, or hot smokers are even weirder. These are geological phenomena that are found on the ocean floor. Cold seeps are horizontal plateaus of growth that resemble coral reefs, but they grow on the ocean floor, in the absence of light. They look like underwater seashores. The microorganisms that build them depend on minerals that are released from fractures in the ocean floor. Thermal vents and hot smokers look like chimneys that spew out warm or hot water. For cold seeps, hydrothermal vents, or hot smokers to be formed, seawater must first enter the ocean crust through fissures, under high pressure. The water is then warmed up (a little bit in the case of cold sweeps, or extensively to produce hot smokers) and will locally escape, enriched with minerals released from the rocks. The water temperature of a hot smoker can reach over 300°C, which is possible at this high depth. No living organisms survive such temperatures, but as one moves away from the center of the plume, a variety of life forms can be found. Crustaceans and worms have specialized to live in the hot, pressurized water, but the highest temperatures where life still persists are reserved for bacteria and especially archaea. They live under such extreme conditions that they are granted a chapter of their own (see Chapter 15).

One novel approach in microbiological research must be mentioned here, as it is frequently applied to marine microbiology. In the past, bacteriology completely depended on the ability to culture bacteria in the laboratory. However, it is estimated that we can only culture one percent of all existing bacterial species. The reasons can be numerous, but all come down to the basics that we cannot provide the organisms with the conditions they need for growth. How to study a bacterium you cannot culture? How to know it is there? Simply, by looking at its DNA.

With the advance of DNA sequencing techniques and PCR amplification (see Chapters 10 and 11), it is now possible to extract DNA from a sample of, in this case, seawater, and sequence all DNA found in it. This approach of sequencing DNA without knowing the organisms it is derived from is called *metagenomics* ("meta" in this case means of a higher kind). A drop of seawater contains a lot of DNA, most of which is of viral origin. There are millions of different viruses in the ocean, and if all these particles were stretched head-to-tail, they would span multiple galaxies. Bacteriologists, not interested in these nonliving DNA packages, remove viruses by filtering the water, as most viruses are smaller than bacteria. The larger protists can also be separated based on their size. When buckets of seawater from the Sargasso Sea (around the Bermudas, off the southeast coast of the United States) were thus treated, scientists from Craig Venter's laboratory started sequencing to see what bacterial DNA was present. They were in for a surprise.

Metagenomic sequencing of Sargasso seawater samples revealed over a million genes, of bacterial and archaeal origin, that had never been encountered before. More than 700 new types of bacteriorhodopsin genes were discovered, which means that far more organisms can use light as an energy source than was previously thought. When all this novel information was added to the public database that stores all sequenced proteins, the size of the database nearly doubled instantly. The sequences obtained for the ribosomal RNA gene (see Chapter 10) identified some species or genera of known organisms that had not been expected to live in the ocean, but even more ribosomal RNA genes were discovered that had to belong to genera or families that had never before been described. Close to 2000 novel and different species are suspected to live in the waters from which the samples were derived, resulting in partly different findings for each bucket drawn from different locations. Now it is recognized that the ocean is truly an unexplored biological environment, whose richness we have only recently started to appreciate. With the vast amounts of water present on our planet, scientists have a long way to go exploring this wet world.

14

Bacteria and Oil

The previous chapter did not mention marine bacteria that are able to clean oil spills, although that subject hit the headlines for several months during the Deepwater Horizon oil spill in the Gulf of Mexico, which started in April 2010. This chapter will make up for the omission, although it is not restricted to marine life, nor will it exclusively deal with oil degradation. Various relationships between bacteria and oil exist, and the most interesting ones will be treated here.

Before addressing which bacteria can feed on oil, it first needs to be defined what oil is. Petroleum ("oil from rock") or crude oil is a mixture of chemical compounds composed of hydrogen and carbon; hence, they are called *hydrocarbons* (not to be confused with carbohydrates, which are sugars). Trace amounts of other elements, mainly nitrogen, oxygen, and sulfur, are also present. The hydrocarbons in natural oil are usually chains of carbon atoms, of variable length from five carbon atoms upwards, to which hydrogen atoms are bound to complete the molecules. The shorter molecules of one to four carbon atoms (methane, ethane, propane, and butane) are gaseous, and some of these gases can be dissolved in the liquid crude oil as long as it is under high pressure in the rocks where it has been formed. When the carbon chains are saturated with hydrogen, they are called alkanes or paraffins, but a considerable fraction of crude oil is composed of either alkenes (molecules in which one or more bonds connecting the carbons are unsaturated double bonds), cyclic alkenes (also called naphthenes), or aromatic hydrocarbons. The latter are cyclic structures of six carbon atoms that are maximally unsaturated, and are better solvable in water than the alkenes. No two batches of crude oil are identical, and even within a single oil well, the composition varies over time

Bacteria: The Benign, the Bad, and the Beautiful, First Edition. Trudy M. Wassenaar.

during extraction. Crude oil containing a high fraction of light hydrocarbons will largely evaporate when it surfaces, whereas a sample with lots of long-chain alkenes ("heavy naphtha") can be viscous to solid, resembling tar and asphalt (both are oil products enriched for long-chain components).

Given this variety in composition, it is hard to predict how oil will behave when it escapes into the sea from the rock in which it had been sealed, either by natural causes or by human intervention. Marine oil seeps occur naturally on a significant scale: together, all natural oil seeps are responsible for 46% of the marine annual load (the other 54% is caused by direct or indirect human activity), as was estimated with surprising accuracy by the National Research Council of the US National Academy of Sciences. All released oil eventually will leave the water column, either by evaporation, deposition, or degradation.

Oil that is lighter than water floats on the surface, spreading out into a thin film, which will cling to the feathers of sea birds, resulting in sad pictures of black oil victims struggling for life. Depending on the oil composition, waves can sometimes emulsify it, as if a giant mixer had whipped it for a salad dressing. The resulting brown mass, nicknamed "chocolate mousse" for its appearance, is both bulkier and heavier than the original oil. Such emulsified oil masses float at deeper water levels, so not all the oil of a spill may be visible at the surface. Alternatively, wave action can result in oil dispersions: small oil droplets that float in the water column without coalescing, like a fine mist. This process can be stimulated by the addition of chemical dispersants. Dispersants are frequently used to combat an oil spill, and they were used in large quantities during the Gulf of Mexico disaster. The reasoning behind this is to reduce the surface oil film, thus potentially allowing less oil to wash ashore or reach sensitive ecosystems, while currents dilute the dispersed oil more effectively. Many oil components are toxic, but toxicity depends on concentration, so the more diluted a spill is, the less toxic it becomes, even if the oil is not removed from the water.

Oil spills are not restricted to the seas, but when oil is released on land (either naturally or by human action), it remains relatively localized, unless it reaches waterways. Whether on land or in the sea, released oil will eventually be degraded or solidified. Biological degradation is a natural but slow process, in which bacteria are key players.

Just like humans, bacteria have discovered that oil has a high energy content: they use oil constituents both as an energy and a carbon source. The saturated alkanes are most quickly degraded; alkenes and smaller sized aromatics are broken down at a slower rate, and multiring aromatics or branched alkenes are a rather tough nut to crack. To fully degrade all components of an oil spill, a microbial community of multiple species is required. This sets research in this area apart from other bacteriological subjects. Most bacteriologists prefer to work with "pure cultures," the offspring of a single bacterial cell that results in a monoculture of identical cells of one species, whereas in Nature bacterial populations are hardly ever monocultures. Scientists studying bioremediation of marine oil spills are used to working with mixed populations of various species.

An oil spill off the Spanish and French coast in 2002 (the result of the wrecked oil tanker Prestige) provided an opportunity to study the natural degradation of a spill microbiologically, and the results have been published since. (A lot of work is currently ongoing on samples taken from the Deepwater Horizon disaster, but those results were not available when this book was written.) The first bacteria to appear at the oil scene were *Alcanivorax* species (a Proteobacteria whose name means alkane eater), of which the well-studied *Alcanivorax borkumensis* is a key example. These degraded the alkanes, after which the more difficult compounds were left for other species to feed on, mostly *Marinobacter*. Over time, a marine oil spill will see the number of *Alcanivorax, Marinobacter, Thallassolituus*, *Cycloclasticus*, and *Oleispira* bacteria rise, as these feast on the hydrocarbons. All of these are Proteobacteria. Many of these bacteria completely depend on oil as a carbon source: without oil, they will not grow. Most of these oil lovers feed on various long-chain alkanes, but some are specialized to live on very particular compounds only. Because oil is diluted quickly in seawater, the choosy bacteria have very efficient means to bind the substrates they prefer so that they can survive on very low concentrations of their favorite hydrocarbons.

Besides these marine "obligate hydrocarbonoclastic bacteria" (literally meaning they have to break hydrocarbons in order to grow), there are also organisms that can use oil as an alternative to a lighter diet. In total, there are over 500 species known of bacteria, fungi, and algae (marine as well as terrestrial) that can degrade oil, but these are metabolically versatile so they can utilize a range of organic substrates, including plant hydrocarbons.

A simple growth test in the laboratory identifies whether oil-eating bacteria are fit to live in the sea: they have to tolerate salt (in most cases, they even need it for growth) at a concentration of approximately 3%. Nearly all oil eaters tolerating or depending on salt are Proteobacteria, but exceptions are a Firmicute, *Planomicrobium alkanoclasticum* (which can degrade alkenes) and the Bacteroidete *Yeosuana aromativorans* (which breaks down cyclic compounds). The *Alcanivorax* species are not exclusively found in the sea: they also have been isolated from soil contaminated with oil, or even from areas near hot springs. Instead of petroleum, those terrestrial bacteria may feed on plant oils.

Since the discovery that bacteria are able to degrade crude oil products, it has been speculated that they could be used to clean up oil spills, by seeding a spill with the right type of bugs, and adding some fertilizer to compensate for nutrients that might limit their growth. The practicalities of such a remedy are not that simple. A lot of research is currently dedicated to understand the various populations of bacteria that actively degrade oil and how we can stimulate their growth, to speed up their cleaning activities. It is still early days, and bioremediation has not been applied to oil spills on a significant scale besides experimental setups. There have been debates whether classical remedies, such as chemical dispersants, help or hinder oil-eating bacteria. The formation of smaller oil droplets increases their relative surface, so that bacteria have better access to their food, but dispersants and their solvents may also have toxic effects. It seems, though, that the dispersants have a stronger positive than negative effect on oil-eating bacteria. An alternative, or additional approach, is to add nutrients that would otherwise limit growth of oil-degrading bacteria, such as nitrogen, phosphorus, or vitamins, which can be applied in lipophilic ("oil-loving") form, to keep the fertilizer in the proximity of the oil, and prevent their dilution into the ocean. All of these bioremediation strategies are still at the design stage, and have not yet been applied to real-life oil spill disasters, so they have not been put to the final test.

Oil-degrading bacteria are not always a bonus. They are active in shallow oil reserves as well, and there they eat the light stuff, to leave heavier oil products behind that are more difficult to extract. There is not much we can do to prevent this. However, when the process occurs anaerobically (in the absence of oxygen), methanogenic bacteria produce methane (CH_4) as a side product. That gas can be extracted to be used as fuel, thus improving the exploration possibilities of these otherwise unexplorable oil fields. Methanogenic bacteria do not live on oil products directly, but depend on the waste products produced by oil-eating bacteria that themselves do not produce methane, providing another example of the interdependence of species within a bacterial community.

After an explored oil well has "dried out," as much as two-thirds of the oil originally present can remain behind, as it is not extractable by classic methods. During extraction, the lighter compounds are favored, resulting in an enrichment of long-chain alkanes and alkenes that are more viscous. Since bacteria are able to degrade these into smaller molecules, resulting in lower viscosity, the application of microbial enhancement of oil recovery has been widely studied. In addition to degradation of heavy oil compounds, some bacteria produce acid as a side product, which dissolves the rock carbonates and improves permeability, again leading to more efficient oil extraction. Although over 400 patents have been issued on this subject, there is still little practical application of these microbial enhancement techniques. One major problem is that bacteria multiply, and in sufficient numbers they plug the wells, a problem that has not been quite resolved yet. Moreover, early experiments were performed with bacteria that would not stand high temperature or pressure, but the more recent discovery of bacteria with the desired properties that survive both may increase their potential as an aid in oil exploration. So far, however, microbial-aided oil extraction has not been practiced outside a laboratory.

The burning of fossil fuel inevitably results in carbon dioxide (CO_2), which is released into the atmosphere. Air contains very little carbon dioxide, so little that we express its concentration not as a percentage but in parts per million (ppm): there are approximately 390 CO_2 molecules in a million other atmospheric molecules. The atmosphere has not remained constant over time, and its CO_2 content varied considerably: its concentration was four to five times higher some 200 million years ago, during the Mesozoic era when dinosaurs ruled the continents. Carbon dioxide is a "greenhouse gas." Its presence insolates like a blanket (or like the glass of a greenhouse, from which the term was taken) and prevents that heat from the sun is being radiated back into space, so that surface temperatures rise when more CO_2 is present in the atmosphere. Other greenhouse gases are methane (which is far more potent than CO_2 but present in even lower amounts in the atmosphere) and even, to a lesser extent, water vapor. As a consequence of the high CO_2 levels in the air, the climate was much warmer during the Mesozoic era, and the ice caps had vanished from both poles, resulting in a much higher sea level than we experience today. Since then, much of that CO_2 has been stored away in newly formed rocks (in the form of carbonates) that have built mountain ridges as a result of tectonic activity. In addition, CO_2 was stored away by the burial of organic matter (plants that had converted carbon dioxide from the air to biomass), which turned to peat and, under the high pressure of the sedimental rocks covering it, to coal and oil. By burning this coal and oil, we release the CO_2 back into the atmosphere faster than natural processes can trap it. The evidence is overwhelming that human activity has resulted in a rise, over the past 150 years, of CO_2 levels from approximately 275 ppm to present-day 390 ppm, an increase of over 40%, mostly as a result of the burning of fossil fuels.

The process of incorporating carbon dioxide into biomatter is called *carbon fixation*. In most cases, this is performed with energy obtained from sunlight during photosynthesis, as was explained in Chapter 13. Plants, algae, and Cyanobacteria are the big players in this game; other autotrophic bacteria (the "self-feeders" that do not need a carbon source other than air) and some archaea can convert carbon into biomass as well, but they may do so with different energy sources than sunlight. The enzyme that binds carbon dioxide from the air and incorporates it into a sugar, from which it can end up in biomass, is called *Rubisco*. This is short for Ribulose-1,5-bisphosphate carboxylase/oxygenase, but we will stick to the short name. Rubisco is possibly the most abundant enzyme on earth. It is the key enzyme in the second part of photosynthesis (after energy from light was captured and stored, as was explained in Chapter 13), which takes place independently of light. The enzyme requires a magnesium ion to function properly. It catalyzes the binding of one carbon dioxide molecule to a sugar made of five carbons (named a ribulose, which, in this case, contains two phosphate groups, hence the "bisphosphate" in the enzyme's name), to produce two molecules of 3-phosphoglycerate. Thus, a sugar of five carbons is split into two smaller molecules, each having three carbon atoms, one of which came from carbon dioxide. Even though that seems more destructive than constructive, 3-phosphoglycerate happens to be a compound that is very easily converted to sugar in the cell. (The cell uses only one of the two produced molecules for building sugars; the other one is regenerated into ribulose bisphosphate, and during this regeneration the hydrogen atoms are used that, as was explained in Chapter 13, were stripped from water during photosynthesis.)

Rubisco is possibly the most important enzyme of the whole biosphere, working in tandem with chlorophyll or related pigments (these are not proteins so they are not called enzymes). It has been suggested that all variants of Rubisco are derived from the enzyme that some archaea still produce; their ancestors must have donated it to the Cyanobacteria, from which it eventually ended up in algae and plants. Maybe because of its ancient origin, it is now an extremely inefficient enzyme, despite its importance and abundance. Rubisco may have evolved at a time when oxygen was absent and during those conditions it may have been efficient. But when oxygen is present, like it is today, the gas competes with its rival CO_2 to bind to the enzyme, hampering carbon fixation. When Rubisco binds oxygen instead of carbon dioxide, it will perform its usual ribulose splicing trick but now it produces only one molecule of the valuable 3-phosphoglycerate, plus a product with the two other carbon atoms; it has fixed oxygen instead of CO_2! Thus, it has both carboxylase (the addition of a carbon atom) and oxygenase (the addition of oxygen) activity, which explains the last part of its long name. Despite this inefficiency, all oil and coal deposits in the world, including the ones we have already burned, were produced with Rubisco.

Society cannot rely forever on the carbon fuels that Rubisco has so industriously produced from carbon dioxide. Not only will our climate change unacceptably because of the released CO_2 if we do but fossil fuels are also finite. If we want to wean society from burning fossil fuels, we will need a number of alternatives. One of these alternative energy sources is to burn plant material that grew recently, at the expense of present-day atmospheric CO_2, so that burning this would not produce a net atmospheric effect. That requires the transformation of plant material into a transportable fuel. Enter biodiesel, the result of the chemical modification of plant oils such as palm, sunflower, or rapeseed oil. Biodiesel is being used as an alternative fuel on a small scale already, although its production is not as efficient as petroleum exploration, because a lot of energy (in the form of fuel) is needed for farming and production, so that the energy yield is modest. Nevertheless, the use of biodiesel results in a decrease of CO_2 release compared to the same energy consumption using fossil fuel. However, as the demand of these plant products increased, serious side effects surfaced. Their production competes with valuable agricultural resources, resulting in a "food versus fuel" competition. Not only will food production suffer but native rain forest is also being burned to make space for palm farms (causing a "forest versus fuel" dilemma). The net result of biodiesel produced from palm oil is that it indirectly produces more CO_2 than it was supposed to prevent, because of the destruction of rain forests.

An alternative fuel produced from plant material is alcohol, resulting from fermentation of plant sugars. In Brazil, many cars are powered by alcohol produced from sugarcane; corn and soy are also a good source for bioalcohol. Nevertheless, concerns are valid that the production of these biofuels cannot be scaled up sufficiently without serious negative consequences on food production, and water and land demand. The development of second-generation biofuel, produced from plant waste products and fast-growing poplar trees, may provide an improvement compared to using eatable harvests or plant oils that compete with tropical rain forest. At best, though, second-generation biofuels may provide a temporary solution only, as eventually their production will also compete for available land, water, and fertilizer resources.

Microorganisms may provide an alternative source to produce either biodiesel or bioalcohol, in the view of some futurists. Algae and bacteria use carbon dioxide from the air and they grow faster than plants. They do not take up as much valuable land, can be harvested year round, and can even grow in wastewater, cleaning this in the process. Bacteria or algae can produce the much-needed long-chain hydrocarbons either naturally, or they can be made to do so as the result of genetic manipulation, whereby all required genes can be added to their existing repertoire, as was explained in Chapter 11. Algae can naturally produce oil in their cells, which they do when their nitrogen source is limiting; they start producing oil as food storage, although they would stop growing as a consequence of their limited diet. Such algal oil can be extracted and chemically modified until it has the desired properties, but the technology has not yet matured outside the laboratory. Alternatively, bacterial communities can produce methane, which is an excellent fuel and a good alternative to oil. It all sounds too good to be true, and at present it still is, but research is ongoing to make microbial biofuel synthesis possible on an industrial scale in due time. Maybe one day microbes can help to saturate our huge energy demand and halt global warming. It would not be the first time they assisted eukaryotes with their small but industrious activities.

15

Extreme Life

This chapter deals less with eubacteria but mostly with archaea, which are true masters of extreme life, as we turn to places on earth where other life forms are unlikely to survive. Harsh conditions for life are formed by extreme temperatures, acidity, pressure, radiation, or salt concentrations. Archaea are true cosmopolitans that receive far less attention than their better-known eubacterial counterparts, an injustice that will be slightly corrected here.

The domain of Archaea, introduced in Chapter 2, can be divided into deep branches that are the equivalent of phyla (some prefer to call these branches "kingdoms"), and those we know most about are the Crenarchaeota and the Euryarchaeota, but there are at least two more phyla. Organisms belonging to the Korarchaeota have never actually been cultured. We know of their existence from metagenomic sequencing, as explained in Chapter 13. A fourth phylum is formed by the Nanoarchaeota, archaea with very small cells. Some textbooks describe archaea as less diverse than eubacteria, but that may simply be a reflection of the difficulty we have to culture them, and their recent discovery: only in 1977 Carl Woese and George Fox recognized them as a separate group, which subsequently got them their own domain within the tree of life. Others describe archaea as always living in extreme conditions, but that is also not correct. As was already discussed in Chapter 13, the oceans are full of archaea, and they also live in the intestines of animals, particularly ruminants, and even in small numbers in our own guts. Nevertheless, talking about extreme life means talking about archaea.

Bacteria: The Benign, the Bad, and the Beautiful, First Edition. Trudy M. Wassenaar.

For some time it was believed that all Crenarchaeota are extremophiles, living under extreme conditions only, such as extremely hot or cold temperatures. Indeed, those species that endure the highest temperatures known are all Crenarchaeota, and many members of this phylum are *thermophiles*, or heat lovers: they occupy niches where temperatures soar. This would apply to the organisms living closest to the center of black smokers at the bottom of the ocean, mentioned in Chapter 13. However, many Crenarchaeota live relatively normal lives in the ocean, and furthermore not all thermophiles are Crenarchaeota, as we will see.

Before we turn to thermophiles that belong to other phyla, we will take one closer look at the Crenarchaeote *Thermofilum pendens*. It was isolated from a solfatara in Iceland, and forms long thin filaments with its elongated cells, but when it divides, the daughter cells start as small round buds at one end of the cell. It can only grow together with another Crenarchaeote, *Thermoproteus tenax*, which produces a number of compounds that *T. pendens* needs but cannot produce all by itself. In fact, when the genome of *T. pendens* was sequenced, it was found that it lacks the genes to produce many of its amino acids and even some of the DNA building blocks. It turns out that *T. pendens* has become completely dependent on other archaea for growth. We have seen loss of genes before, but in contrast with the insect endosymbionts discussed in Chapter 8, *T. pendens* does not have a small genome; it has just lost a number of genes that few organisms can effort to lose, as it makes them dependent on their surroundings. Terrestrial hot springs are frequent homes for extreme thermophiles. The beautiful colors of geysers and hot springs, such as can be seen in Yellowstone National Park (United States), are mainly produced by thermophilic Crenarchaeota, which also like to live on rock surfaces of volcanoes, deserts, or even in the chimneys of power plants. Most of these heat-loving species cannot multiply at temperatures below 70°C, though they may survive such "cold" conditions and wait till it warms up again.

For thermophiles that live close to a hot smoker at the bottom of an ocean, life is demanding, as they have to survive and grow at extremely high temperatures and immense pressure, due to the depth. Such conditions do not present challenges for the cell membrane, as long as the pressure inside the cell equals that outside. The high pressure and high temperature inside the cell does, however, require that the enzymes perform their chemical reactions under conditions very different from that of more usual life forms. It is believed that resistance to both high pressure and high temperature is enabled by a single strategy. Studies with *Methanococcus*, *Thermococcus*, or *Pyrococcus* species (which are all Euryarchaeota, to give an example of thermophiles that are not Crenarchaeota) indicate that their enzymes are pressure stable, and enzymes from high-pressure organisms are also thermostable. The trick is that the enzymes form structures that are chemically extremely robust.

High temperatures can also be combined with extreme acidity and these conditions are favored by *thermoacidophiles* that, as the name suggests, need a hot as well as acidic environment. For instance, in the hot springs in Beppu, Japan, sulfuric acid close to boiling temperature is the home of a number of thermoacidophiles. These are a large group within the Crenarchaeota with *Sulfolobus* species as well-studied examples, or organisms from a genus with the stinging name *Picrophilus* (an Euryarchaeote) that can survive in extreme acidic environments. However, not all acid-loving microbes are archaea; "acidophile" eubacteria also exist but they do not like it so extreme. An extremely acidic crater lake of ambient temperature on the island of Java (Indonesia) is full of archaeal acidophiles, but the river that is fed by this lake, which is slightly less acidic, also harbors acidophilic eubacteria.

The opposite of an acidic environment is an alkaline solution, and alkaline springs or lakes are inhabited by *alkaliphiles*. The discovery of alkaliphiles is relatively recent. A lot of research was done in Japan, for historical reasons. The indigo industry was thriving in classical Japan during the Edo period (the seventeenth and most of the eighteenth century), when silk was expensive and cotton was commonly used for clothing. Indigo was one of the few dyes that could color cotton, and even today jeans are blue because of indigo. It is produced chemically in present times, whereas traditionally the dye was made using the Indigo plant (*Indigofera tinctoria*). *Indigofera* plants do not actually contain indigo and they are not blue, as indigo has to be produced by chemical reduction of the plant's indole compounds. This reaction takes place under alkaline conditions. The leaves of the plants were fermented in the presence of sodium carbonate, which creates an alkaline solution, and alkaline-loving bacteria were responsible for the chemical reactions that produced the blue dye.

A number of different alkaliphilic bacteria were probably present in the classic indigo vats, such as alkaliphilic members of *Clostridium*, *Pseudomonas*, *Cyanobacterium*, *Micrococcus*, or *Bacillus* (all of these are Eubacteria, of various phyla). Enzymes isolated from the alkaliphilic *Bacillus* species are used nowadays in detergents, as they have to work under alkaline conditions. Similarly, hand soap is moderately alkaline and a piece of used hand soap is home to a number of alkaliphiles, which can easily be cultivated in the laboratory. Soap is not designed to kill bacteria, but to remove dirt from your hands more easily than water alone would do. There is no reason to worry, but the knowledge that soap is far from sterile puts the hand-washing advocacy of hygiene-lovers into perspective. Changing to antibacterial soap should not be considered an improvement, as it was demonstrated that such soaps do not clean hands, or prevent the spread of disease, any better than normal soap, while the disadvantage of antibacterial products for household use was already outlined in Chapter 12.

High concentrations of salt are no problem for a group of organisms described as *halophiles*. The more extreme salt lovers are all archaea. They live in the Dead Sea (whose name is obviously incorrect) or in salt lakes such as the Great Salt Lake in Utah (United States). Coastal lagoons and man-made salterns also provide a suitable living. Halophiles frequently produce pigments, like the thermophiles do, which can color salt lakes reddish or pink. Extreme halophiles thrive in water containing up to 30% sodium chloride salt. Organisms living in normal conditions take care not to swell up by intake of water, and must prevent losing their minerals, which would diffuse out of their cells if they did not keep pumping ions back in. Halophiles, in contrast, meet the opposite challenge of having to prevent salt ions entering their cells spontaneously, or water leaving their cells. Some do this by stocking up high concentrations of potassium salts, which stops spontaneous water loss. Other species have developed an alternative strategy: they keep their cell content low of salt by accumulation of uncharged, soluble organic matter, such as glycine betaine. This serves as a protectant, not only against salt but also against heat, freezing and thawing, or desiccation. It provides these archaea with the means to live in a wider range of environments. In addition, they also have dedicated ion pumps in their cell membrane which pump ions out, not in. As one can imagine, extremophiles using this strategy can endure quite a bit of discomfort. *Haloalkaliphiles* combine the challenge of high salt with alkaline conditions, and these are mostly archaea that produce bright pink pigments. They can be found in natural lakes rich in salts and highly alkaline, for instance, Wadi Natrun in Egypt.

Halophiles can survive extreme conditions, for an extremely long time. Live archaea (*Halococcus* species, Euryarchaeota) have been isolated from salt deposits that originated during the Permo-Triassic age. They must have lived (and survived) there for about 250 million years, and the characterization of three isolates, originating from ancient salt deposits in Germany, England, and Austria, showed they were all very similar. Maybe these deposits were formed in an ancient hyper-saline shallow sea or lagoon that was inhabited by these organisms. The lagoon must have accumulated more and more salt, and the archaea hid and survived in the salt deposits at various locations, to be discovered by microbiologists after quite some time.

Life can also thrive in extremely low temperatures. The label *psychrophiles* has been coined for bacteria that like it cold ("psykhros" is Greek for cold). Some psychrophiles are used in the production of ice cream, as their enzymes perform optimally at freezing temperatures. The bacteria living in the deep ocean are also psychrophilic, as the water will not be warmer than 5°C. But Korarchaeota living in the ice of the poles can withstand temperatures much lower than that, and some combine the cold with additional conditions that would have made life miserable if they had not adapted. One adaptation is to produce antifreeze proteins inside the cells, a trick that arctic fish were shown to master before it was discovered in bacteria, too. It is likely, though, that bacteria produced antifreeze before fish did. The order in which properties are discovered should not be taken to mean things had evolved in that order as well.

Some bacteria live between a rock and a hard place. Polar deserts are very cold—the climatologic term *desert* indicates lack of precipitation, and that can occur in a cold as well as in a hot climate. Moreover, cold air contains very little water. The McMurdo Dry Valleys in Victoria Land, Antarctica, are hyperarid deserts with a permafrost soil covered by a gravel pavement. Bacteria of various phyla can be found in the cracks, fissures, and pores of the rocks, but surprisingly, archaea appear to be absent. They seem unable to cope with the lack of water. Even the frozen soil is colonized, mainly by bacteria of the genus *Deinococcus* (their phylum Deinococcus-Thermus is partly named after them). These organisms are not only resistant to desiccation but also to irradiation. *Deinococcus radiodurans*, which was originally isolated from canned food that had been sterilized by radiation, is extremely resistant to radioactivity. These bacteria can survive a radioactive dose 3000 times stronger than a person can survive, thanks to the remarkable ease with which they repair DNA damage.

Life in the deep seas was treated in a previous chapter, but what about life in the deep earth? It is there, in the form of bacteria living in deep groundwater reservoirs that are completely sealed off from life at the surface. It means they are on their own, and need to produce all their biomolecules from inorganic material. Sunlight can obviously not be used as a source of energy. Instead, these bacteria may use energy released from the oxidation of iron, or even from radiation energy. A community of Firmicutes lives at a depth of 2.8 km below the surface of the Witwatersrand basin near Johannesburg (South Africa). They seem to be happy living on their own, as all bacteria detected in fracture fluids in these deep rocks were of the same species, provisionally named *Desulforudis audaxviator* (this name still has "Candidatus" status, meaning it is not yet officially accepted by an international board of name-giving taxonomists). These bacteria can make all their biomolecules from dissolved inorganic salts, which they have to, since there are no other living forms down there that provide organic compounds for them. They are a rare example of a naturally occurring bacterial monoculture.

Even the ocean crust, deep under the water column, harbors life. A recent drilling of the ocean floor of the Atlantic (some 10 km below the water surface) indicated the presence of Proteobacteria in rocks collected from 0.4 m to over 1 km deep below the sea floor. These small populations (there were not that many bacteria present) are related to the hydrocarbon degraders we have met in the previous chapter. Their carbon and energy source is most likely hydrocarbons, whereas these are not formed from plant material, the way oil present in shallow-water oil fields was formed. Instead, their oily food is of geological origin, formed without the help of life.

Sometimes, however, bacteria get lost and end up in places where they do not belong. What else to think of thermophiles found in arctic ice, or alkaliphilic organisms isolated from acidic soil? These poor things got there by accident, and may survive, but they are unlikely to multiply. When bacterial spores of thermophiles, isolated from subzero sediments off the island of Spitsbergen (within the Arctic Circle), were incubated at 60°C, they sprung to life. It demonstrates how right Martinus W. Beijerinck (1851–1931), a Dutch microbiologist, botanist, and founder of virology, was when he commented: "everything is everywhere, but the environment selects." Bacteria and archaea may reach every niche on earth, but the local environment dictates which species can grow there.

16

Record Holders

Record holders usually receive more attention than the group that collectively is most common, the modal. Since our knowledge of the bacterial world is skewed,

Bacteria: The Benign, the Bad, and the Beautiful, First Edition. Trudy M. Wassenaar.

a lot remains unknown to this day, both about the modal and the record holders of microbial communities. Records are bound to be broken as new discoveries are made, and the modal may shift as we learn more about what is common. Nevertheless, here is an attempt to present the currently known winners from a variety of contests.

Bacteria are usually small, so we will start with size. Being big is a problem for bacteria, as they need to take up their food through passive diffusion, and the bigger a body is, the smaller is its surface to volume ratio. In other words, a small bacterium has a larger surface, relatively speaking, which allows more efficient traffic of solutes. An example of an exceptionally big bacterium is *Epulopiscium fishelsoni*, a Firmicute isolated from the guts of tropical fish; it must be literally living in its food, as it can grow to 0.6-mm-long rods, and must diffuse its food despite its unfavorable dimensions. However, it is not yet breaking the records.

The biggest bacterium known to date is *Thiomargarita namibiensis*, a Gram-negative Proteobacterium that lives in mud on the sea floor off the Namibian coast. These bacterial giants measure 0.75 mm, making their individual cells easily visible to the naked eye. They grow in chains and look like little white beads of a pearl string against the dark mud, which gave them their name (margaron is Greek for pearl). These bacteria need sulfur, and can grow so big since they store their food inside their own cells, in specialized sacs called *vacuoles*. Similar bacteria were found off the coast of northern Chile, where immense mats of white hairlike growth were discovered to be made of giant sulfur-eating bacteria. They were one of the many discoveries of the Census of Marine Life project, which aims to catalogue all ocean life. The findings published so far have already revolutionized our view on biological diversity. It is now estimated that microbes may make up 90% of all marine biomass, and there may be 10 times as many species as previously suspected.

The smallest bacteria are easily 15,000 times smaller than these microbial giants. The miniatures are generally described as "nanobacteria" but that is not a taxonomic term; it only refers to their size. Nanobacteria measure at most a few ten-thousandth of a millimeter (0.2 μm, which is the same as 200 nm) and can be either eubacteria or archaea. The archaeal *Thermodiscus maritimus*, a Crenarcaheote which lives in the ocean, looks like a microscopic flying saucer; their disk-shaped cells have a diameter of only 0.2 μm, and a thickness of 0.1 μm. A number of nanoscale archaea that turned out to be related to each other were combined into a separate phylum, called Nanoarchaeota, which form a separate deep branch of the Archaea domain. The first discovered Nanoarchaeota grew like little warts on the surface of cells of a normal-sized archaea, belonging to the *Ignicoccus* genus (a Crenarchaeote). The two are probably symbionts, as one would not grow without the other. Their partnership was discovered as a result of a diving excursion to explore a hydrothermal system north of Iceland, in water 100 m deep. *Ignicoccus hospitalis* and its tiny friend *Nanoarchaeum equitans* like a temperature of around 90°C.

True bacteria (eubacteria) can also be very small, for instance, *Mycoplasma* (not to be confused with *Mycobacterium* species). *Mycoplasma* are neither Gram-positive nor Gram-negative, but they seem to have started as the first, and subsequently lost their cell wall. They are granted their own phylum called Mollicutes, which means soft skin. These bacteria live as parasites inside plant, insect, and human or animal cells. In humans and animals, they prefer the mucosal lining of respiratory or urogenital organs, or they infect the eyes, depending on the species. *Mycoplasma* cells are so small that 200 million of them would occupy the same volume as one *Thiomargarita* cell. How does that compare to the animal world? The smallest insect (a male parasitic wasp, about 0.14 mm long) is 3 million times smaller than the biggest mammal (the blue whale, 30 m), but the size of their animal cells is of the same order of magnitude. Bacteria, consisting of one cell only, owe their variation in size solely to variation in cell size.

More about size

In general, bacterial cells are smaller than eukaryotic cells, and viruses are smaller still. Eukaryotic cells usually measure between 100 and 10 μm (a μm is thousandth of a millimeter), while bacteria are typically 10 times smaller. Viruses are again 10 to 100 times smaller than bacteria. However, some observations do not fit these general rules:

- The largest bacterial cells measure 0.75 mm, which would be large even for a eukaryotic cell. Human cells typically measure 10 to 20 μm, although our largest cells are of size 0.13 mm (equal to 130 μm), and the longest brain cells can reach from head to toe. The largest eukaryotic single cell is the ostrich egg, which is about 15 cm (6 in).
- The smallest eukaryotic cells belong to single-cellular algae. The cells of *Ostreococcus tauri* are the size of small bacteria with a diameter of only 1 μm.
- The smallest bacterial cells span only 0.2 μm, which is within the range of a typically sized virus. Viruses come in many shapes and sizes, usually measuring between 0.3 μm and a few hundredth of a micrometer only.
- The largest virus known to date is 0.75 μm in diameter, nearly four times bigger than the smallest bacteria and visible under a light microscope. After the discovery of giant mimivirus, which parasitizes on a soil amoeba and whose genome contains 400 genes, the record was broken by the largest marine virus so far discovered that parasitizes on zooplankton: it contains more than 500 genes, equal to the gene content of bacteria with a small genome. These large viruses exist at the border of life and nonlife.

Very small cells do not have much space for their DNA, so it is no surprise that small cells have small genomes. That of *Mycoplasma* is only 600 kb, carrying a bit over 500 genes, which is possible because they live as parasites, so that a number of their cellular processes are taken care of by their host. But the small genomes of *Mycoplasma* are not the record. The symbiotic (or parasitic, it is hard to tell how much its archaean host suffers from its presence) *Nanoarchaeum equitans* only has 490 kb of DNA and this is beaten again by some of the eubacterial endosymbionts. The Proteobacterium *Carsonella ruddii* lives as an endosymbiont in cells of psyllids (plant lice) with only 160 kb of genetic material at its disposal. It has a mere 180 genes in total. In Chapter 6 it was already discussed whether such cells can still be considered alive, as they have lost so many genes that they depend on their host cell for many essential steps of their metabolism. At the other end of the size scale, though, giant bacteria do not have exceptionally large genomes.

The largest bacterial genomes to date size more than 10,000 kb (10 Mb) and contain about 50 times more genes than *C. ruddii*. These record holders are *Solibacter usitatus* (a soil bacterium isolated from pastureland in Australia), or a Myxobacterium called *Sorangium cellulosum*. These have large genomes, for a bacterium, but normal-sized cells. They need so many genes because they can survive and grow under many different conditions, whereas the endosymbionts have a restricted lifestyle. By the way, a human cell does not have that large a genome, either. The size of our genome is easily beaten by amoeba, some of which have the largest genomes known so far, containing about 100 times more DNA than our cells do. In eukaryotes, not all that DNA codes for genes, as lots of DNA sequences are repeats of short "words" for which we cannot recognize a function. Bacteria are usually more economic and use most of their DNA for genes, so that a large bacterial genome implies a large number of genes present.

A DNA molecule (and thus a genome) is made of four bases, abbreviated with the letters A, G, C and T. The DNA molecule is built of a structure resembling a spiral staircase, or rope ladder (the famous double helix, first described by Watson and Crick) where the bases form the steps or rungs. One DNA strand produces the halves of all steps, the other DNA strand provides the other halves of these steps. Every step thus comprises a pair consisting of two bases, one from each strand. A pair of an A and a T forms one kind of step. Since this step is composed of two different ends, we can reverse this to have a step of T paired with A. The other kind of step is produced when G pairs with C (or C with G). This gives the four different steps, of which all DNA ladders are built. At any position in a double helix DNA molecule, one will find an A on one strand where a T is present on the other strand. Similarly, a G will always match with a C on the other, or "complementary" strand. That is not to say that DNA always contains equal numbers of each base. Although at one position a C always needs a G to form a base pair, and a T always needs an A, one could imagine a ladder with lots of A-T steps and only a few G-C steps. Indeed, there are bacteria that have very imbalanced DNA composition: their genomes are far richer in G-C bases than in A-T bases, or the other way round.

The record holders of AT-rich DNA are, again, the endosymbionts. *C. ruddii*, the current winner of the contest "Who Has the Smallest Genome," also has the highest AT content: approximately 84% of its DNA is made of AT base pairs. The *Mycoplasma* species also contain extremely AT-rich DNA. One species of this genus, *Mycoplasma pneumoniae*, can cause pneumonia but *Mycoplasma genitalium* causes a sexually transmitted disease in some persons, although a large number of infected individuals do not display any symptoms. It is not known why endosymbiotic bacteria frequently contain DNA so rich in A and T.

Bacteria can also have DNA with lots of Gs and Cs. Examples are soil bacteria, such as the social Myxobacteria, which can have DNA with a GC content of up to 74%. The reason for this vast variation in base content, from very high AT to very high GC, is not exactly known, but it has evolutionary implications. Imagine what would happen if an AT-rich bacterium were to eat a bit of DNA left over from a GC-rich organism. The recipient would find it difficult to use the genes coded on it, as the base composition would be so different from all its other genes that it would hamper its protein synthesis. The components for protein synthesis are optimized for the base composition of the cell's DNA, and although slight variation between genes is allowed, a gene with a very different base composition will not be efficiently translated into protein. Indeed, the DNA of bacteria that share the same ecosystem tends to vary less with regard to base composition than the DNA of bacteria living in very different ecosystems. Whether this is because bacteria can share a bit of their DNA when living together, or they live together because they can potentially share DNA, is a chicken-and-egg problem.

Since bacteria can grow rapidly, the fastest growing species deserves to be honored here. The winner probably is *Clostridium perfringens*, which we have seen in Chapter 9 is a common cause of food poisoning. These Gram-positive bacteria are able to divide in a mere 6.5 min. However, the conditions under which a population can multiply influence their growth rate. Temperature has a large effect. *C. perfringens* likes temperatures between 36 and 46°C, with the fastest growth observed at 43°C. The nutrients available for the cells also have an effect: the more complex molecules they can feed on, the fewer of these biomolecules they have to produce themselves, so they can put all their energy in growth. Unfortunately, ground beef provides quite a rich diet for *C. perfringens*, and on this food they can multiply in 7 to 9 min, depending on the bacterial strain that was tested. This rapid growth poses a serious problem to the food industry, as large quantities of cooling foods that happened to be infected can stay at permissive temperatures long enough for the bacteria to reach critical numbers.

The prize for the most cruel bacterium possibly goes to *Bdellovibrio bacteriovorus*, which lives in soil or river water. These Gram-negative Proteobacteria whose names means "bacteria eater" are very rapid swimmers. They can cover a distance of 100 times their body length in a second, by means of their single flagellum (see Chapter 4). They use this speed to collide with other Gram-negative bacteria, which they enter by crashing through the outer membrane. *B. bacterio- vorus* then stays in the space between the outer and inner membrane of its prey. Membranes are slightly flexible, so although the space between the two membranes of Gram-negative bacteria is usually tight, space can be made for the unwanted guest. Then the *Bdellovibrio* starts eating. It eats the complete content of the cell it has penetrated from the inside. The *Bdellovibrio* bacteria start dividing inside the dying cell and within one or two days, the dead cell bursts and the preying bacteria are released, to start swimming in search of a new victim.

If bacteria were intelligent, the winners of an IQ contest would be species that can count. Some bacteria are able to estimate the size of their population, and, depending on how many of their likes are out there, regulate their numbers. This phenomenon is seen in bacteria that do not only grow as lone wanderers, but that can also live in close communities, the microbial equivalent of a city. Although microbiologists usually study free-floating bacteria, from which they can produce single colonies on an agar plate, many bacteria can switch lifestyle to form so-called *biofilms*, in which cells from single or multiple species grow closely together. Biofilms are densely packed communities of bacteria that grow on surfaces and surround themselves with a slime of sugar polymers. Bacteria in a biofilm behave as coordinated and cooperative groups, analogous to cells in multicellular organisms. Bacteria that form biofilms can sense how many they are, by a process called *quorum sensing*. Depending on what their quorum sensing signals tell them, they will continue dividing, or stop growing and produce sticky slime. Some species in a biofilm produce slime when they have reached high numbers, but other species, in fact, stop producing slime when there are enough of them. Quorum sensing not only regulates the production of biofilm slime but also whether cells attach, activate their bioluminescence proteins, or start to form spores, depending on their capacities. It is now recognized that bacteria can be highly social organisms. The slime that covers the multiple layers of bacteria in a biofilm makes them insensitive to antibiotics, as the drugs cannot reach the cells. This is why biofilms that form in the body are hard to treat, for instance, in chronic otitis media, or biofilms that grow on implanted shunts or catheders. Biofilms also form inside water pipes, and can grow to dimensions that hamper water flow.

Talking about water pipes, the prize for the most adaptive opportunist will go to Gram-negative *Legionella pneumophila*. These bacteria were discovered in 1976 as the cause of an unusual outbreak of pneumonia. In July of that year, 221 attendants to the fifty-eighth annual meeting of the American Legion's Pennsylvania Chapter, who met in Philadelphia (United States), fell ill. Of the 221 persons who got sick, 72 died. The lung infection that they suffered from is known since as Legionnaires' disease and the bacteria that were discovered as the cause were named *Legionella pneumophila*. These Proteobacteria happen to grow in hot-water pipes, and fine mists of aerosols with a high bacterial load can be produced by showers and Jacuzzis. The showers of the hotel where the legionnaires stayed were the cause of the outbreak. Since its discovery, it was recognized that *L. pneumophila* is quite common in hot water pipes or cooling towers. Sadly, the disease has remained with us ever since it was discovered. In 2000, a large outbreak in the Spanish city of Murcia, involving at least 745 cases and one death, was caused by bacteria that had multiplied in a cooling tower and were spread over the town by the wind. By mapping the cases on a town map, one could see where the "plume" that had caused so many illnesses had originated. An even more unusual outbreak happened in the previous year in the Netherlands. During an indoor flower show, which attracted 77,000 visitors, an exhibit area displayed a hot-water whirlpool spa. Unknown to the organizers, this spa was fed with water that happened to be contaminated with *L. pneumoniae*, and the whirlpool effectively produced aerosols that spread the bacteria amongst the visitors. At least 188 persons fell ill, and 21 died. With an apparent preference to live in hot-water pipes, one can only wonder where *L. pneumophila* lived before man invented hot showers.

One bacterial species is the winner when it comes to difficult names. It is *Aggregatibacter actinomycetemcomitans*, record holder of a tongue twister. It used to be called *Actinobacillus actinomycetemcomitans*, which was equally bad. The species belongs to the Gram-negative Proteobacteria and lives in the mouth of humans and animals, where it can cause periodontal disease. To give a Ph. D. student this species to work on is kind of cruel. One day, the poor student will have to present his or her findings at a scientific meeting, in front of a large audience, with shivering knees and a dry mouth. Only ten minutes are typically available for an oral presentation, in which research of at least a year's work has to be cramped. But a considerable proportion of the available time will be lost with pronouncing the species' name alone. A slow speaker would be advised to work with short-mouthed *Frankia alni*, a symbiont to the green alder tree.

The beauty contest amongst bacteria is a difficult one, because, as the saying goes, beauty is in the eye of the beholder. Under a microscope, the members of the Spirochaete phylum, such as *Leptospira*, with their long corkscrew bodies, certainly look neater than plump *Enterococcus* (a Firmicute). Despite these looks, the *Leptospira* species are quite nasty: leptospirosis is a common zoonosis (a disease humans contract from animals), which people catch after contact with surface water contaminated with infected rodent's urine. It can cause large outbreaks during floods. But no matter how pretty their cells are, the colonies *Leptospira* form on agar plates that are visible with the naked eye are rather dull. Bacterial growth is often not as spectacular as the patterns that molds and fungi can produce, although the colonies of multicolored *Streptomyces* can certainly be appealing to the eye. However, the beauty queen is *Paenibacillus dendritiformis*, a common soil inhabitant that is a member of the Firmicutes, and was previously mistaken as *Bacillus subtilis*. The bacteria are motile by propulsion, as was explained in Chapter 4: they secrete a slime to be able to move over solid surfaces. These bacteria may actually win a bunch of medals, as they are also very social and seem to be quite clever. The cells of *P. dendritiformis* can sense food, learn from experience, and remember where they are going. Collectively, the cells of a colony take decisions as if they are highly organized.

Does that sound a bit over the top for a "simple" bacterium? The colored illustrations show what these little wonders are capable of. The pictures belong to a series of remarkable patterns that *P. dendritiformis*, and its cousin *Paenibacillus vortex* can form when grown on agar plates under different conditions. On normal agar plates, mostly boring grayish colonies form, similar to those of so many other bacteria. But starve them by giving them very little nutrients, and the most beautiful fractal patterns will appear. The bacteria have evolved to survive difficult conditions by developing cooperation and communication skills. Blessed with these skills, the bacteria behave as if they were taught esthetics by an artist. How bacteria and artists interact is further explored in Chapter 17.

17

Bacteria and Art

The colored illustrations in this book are examples of pieces of art produced or inspired by bacteria. The first six are produced by Eshel Ben-Jacob. The patterns that *Paenibacillus dendritiformis* and its close relative *Paenibacillus vortex* produce on agar plates when grown under food-limiting conditions at first drew his attention because of their physical properties. Being a physicist, Ben-Jacob investigated what parameters caused the formation of particular patterns and how these could be mathematically described. But he was also inspired by the esthetics of these little wonders of nature, so he experimented with growth conditions to produce the most beautiful patterns. The artist used the bacteria and varied their growth medium to produce these compositions, the formation of which depended on bacterial choices for the direction in which to move and grow. He then added the colors to the photographs artificially. For these experiments, the artist inoculates an agar plate with a small droplet containing between 10,000 and 100,000 bacteria. After several hours of bacteria multiplying within the center dot, food becomes locally scarce. The colony sends out branches that spread on the surface of the plate in search for food. As the bacteria continue to grow, they push the branches forward by moving on the lubricating slime that they produce.

The other colored illustrations are paintings by Karoly Farkas, who finds inspiration in micrographs of bacteria. Bacteria under a microscope are colorless, so the coloristic choices are again an artistic touch to the compositions. The variety in

Bacteria: The Benign, the Bad, and the Beautiful, First Edition. Trudy M. Wassenaar.

shapes fascinates Farkas, and to him, bacteria are "organic," connecting the invisible with the world around us. He would not hesitate to call bacteria beautiful. However, not every artist would make that association.

Bacteria have a complex relationship with art. On one hand, bacteria produce beautiful patterns that can be artful in composition and color; on the other, pieces of art do not mix well with bacteria, as the latter can be quite destructive. In addition, bacteria have prevented the creation of many imaginary pieces of art. Numerous music compositions, books, paintings, or sculptures never saw the light of day because a talented artist died prematurely of an infectious disease.

A number of famous artists died an early death because of bacterial infections. Artists were probably not more likely to suffer from infections than people with other professions, but their causes of death had a better chance of being recorded and remembered. A number of famous composers are known to have died of an infection. Franz Schubert (Austria, 1797–1828), best known for his "Lieder" (German songs for voice and piano), suffered from syphilis (caused by sexually transmitted *Treponema pallidum*), which he discovered when he was only 25 years old. The disease not only weakened him but also caused mood swings, and in times he was depressed, which is reflected in some of his later works. Syphilis had severely weakened him when, according to history, he died at an age of 32 from typhoid fever (caused by *Salmonella typhi*), after being on the sickbed for a short time. This may have been the case, but it is more likely that his physician changed his cause of death to spare his family the shame that the diagnosis of a sexually transmitted disease would have brought upon them.

Robert Schumann (Germany, 1810–1856), who mainly composed for the piano, also suffered from syphilis and sadly experienced the final, neurological stage of this disease that occurs in approximately 25% of untreated infections: the bacteria had entered his brain and caused hallucinations and early dementia. He attempted suicide, after which he was confined to a mental institution, where he died of his disease two years later at the age of 46.

Pjotr Iljitch Tchaikovski (Russia, 1840–1893) died nine days after the premiere of his sixth symphony, now known as *Pathetique*, which at that time was not so well received. His works include the ballets Swan Lake, Sleeping Beauty, and The Nutcracker; three piano and a violin concerto; and six symphonies. Occasionally he was depressed and suicidal, and rumors about a homosexual relationship troubled his reputation. His cause of death is recorded as cholera, and this infection may have been accidental. Or maybe it was intentional, although it has also been suggested that he killed himself by taking arsenic poison.

Even when not by personal suffering, infections could still influence an artist's works. Edvard Munch (Norway, 1863–1944), to give an example of a painter, lost both his mother and sister to tuberculosis (the disease caused by *Mycobacterium tuberculosis*). He painted "Sick Child" in memory of his beloved sister. His works breathe death, suffering, and despair, and his most famous work, "The Scream," seems to sum up all agony of life.

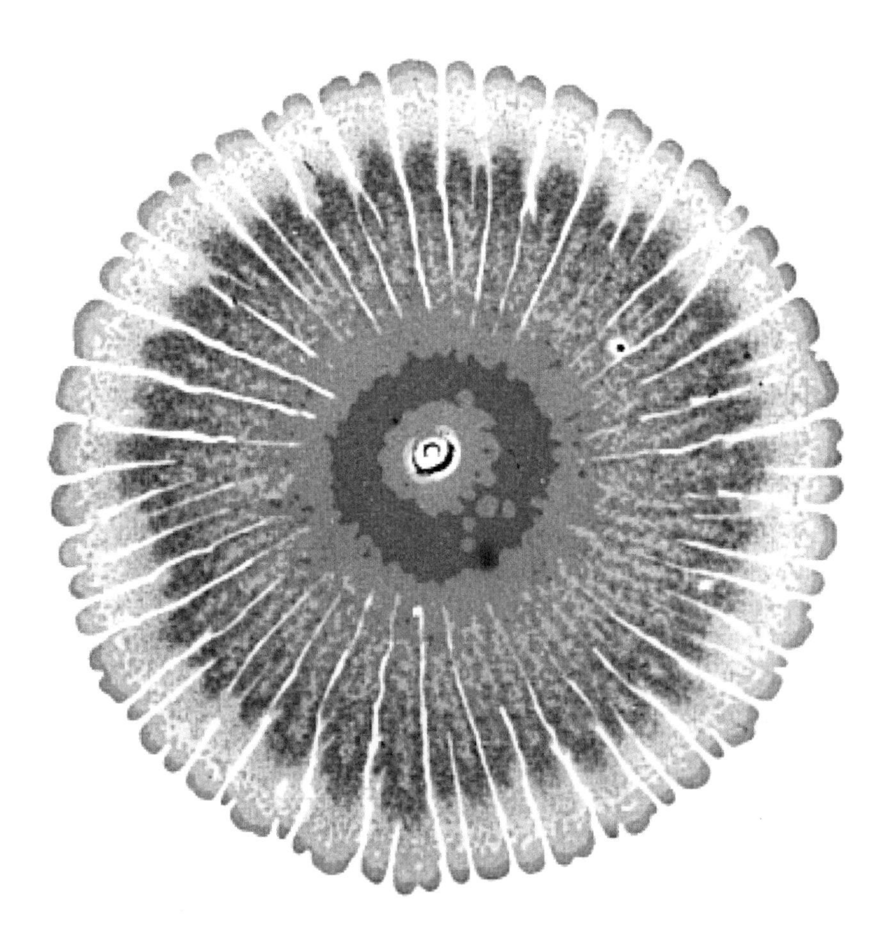

This colony of *Paenibacillus dendritiformis* was grown for 8 days on an agar plate that provided a lot of moisture and a strong, solid support, but very little food, containing only 0.5% gram/liter peptone. From the center of the plate, where the bacteria had been seeded, the colony spread in a beautiful wheel of radiant growth.

Here, *Paenibacillus dendritiformis* bacteria were grown on poor nutrients on a hard surface. Their medium contained 2 g/L peptone and 1.5% agar. The dot in the center is where bacteria were inoculated. The colony diameter is 6 cm.

By providing *Paenibacillus dendritiformis* with a little more food, the colony changes shape. These cells were grown with 5 g/L peptone on a softer surface, produced with 1% agar. The colony width is 4 cm.

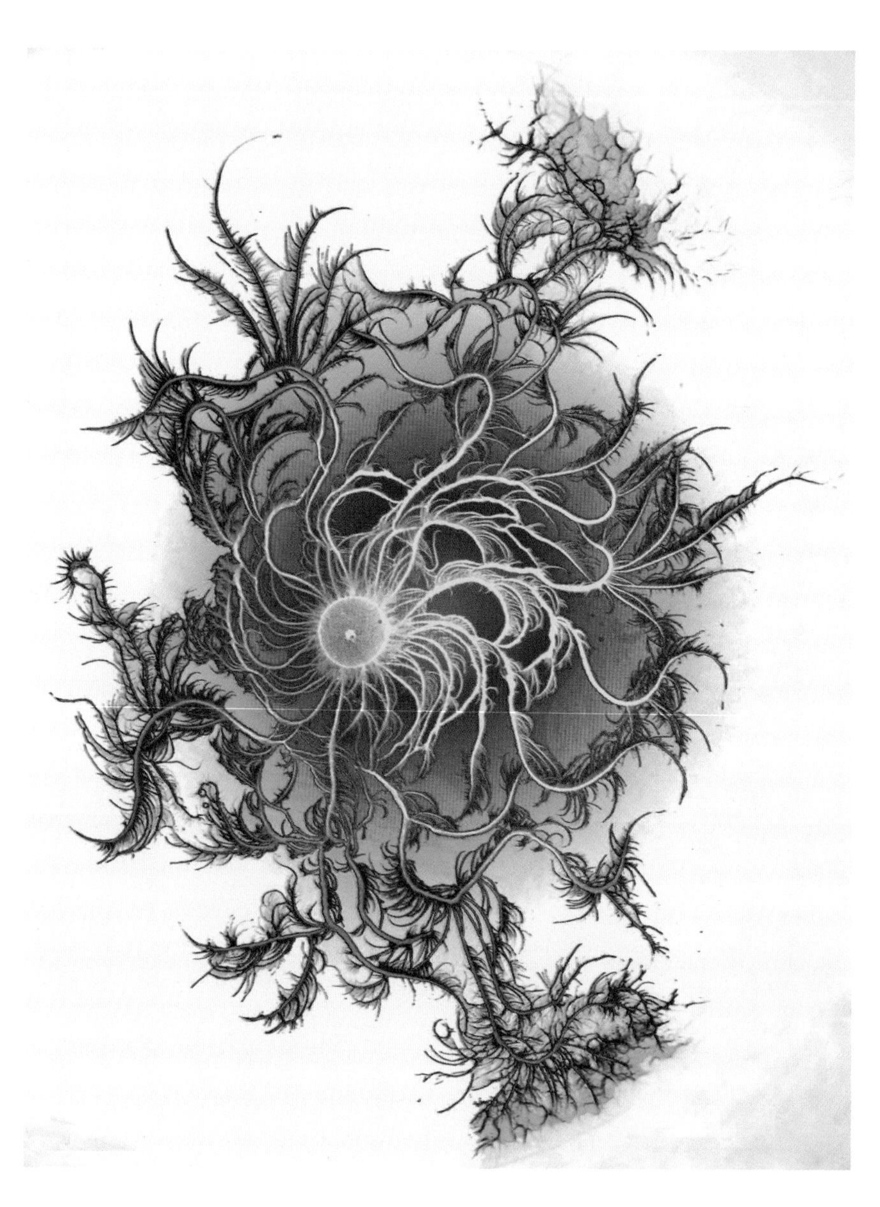

The same *Paenibacillus dendritiformis* bacteria as in the previous illustrations were grown here on poor nutrients and a soft surface (their medium contained 2 g/L peptone and 1% agar). These bacteria belong to the so-called branching morphotype.

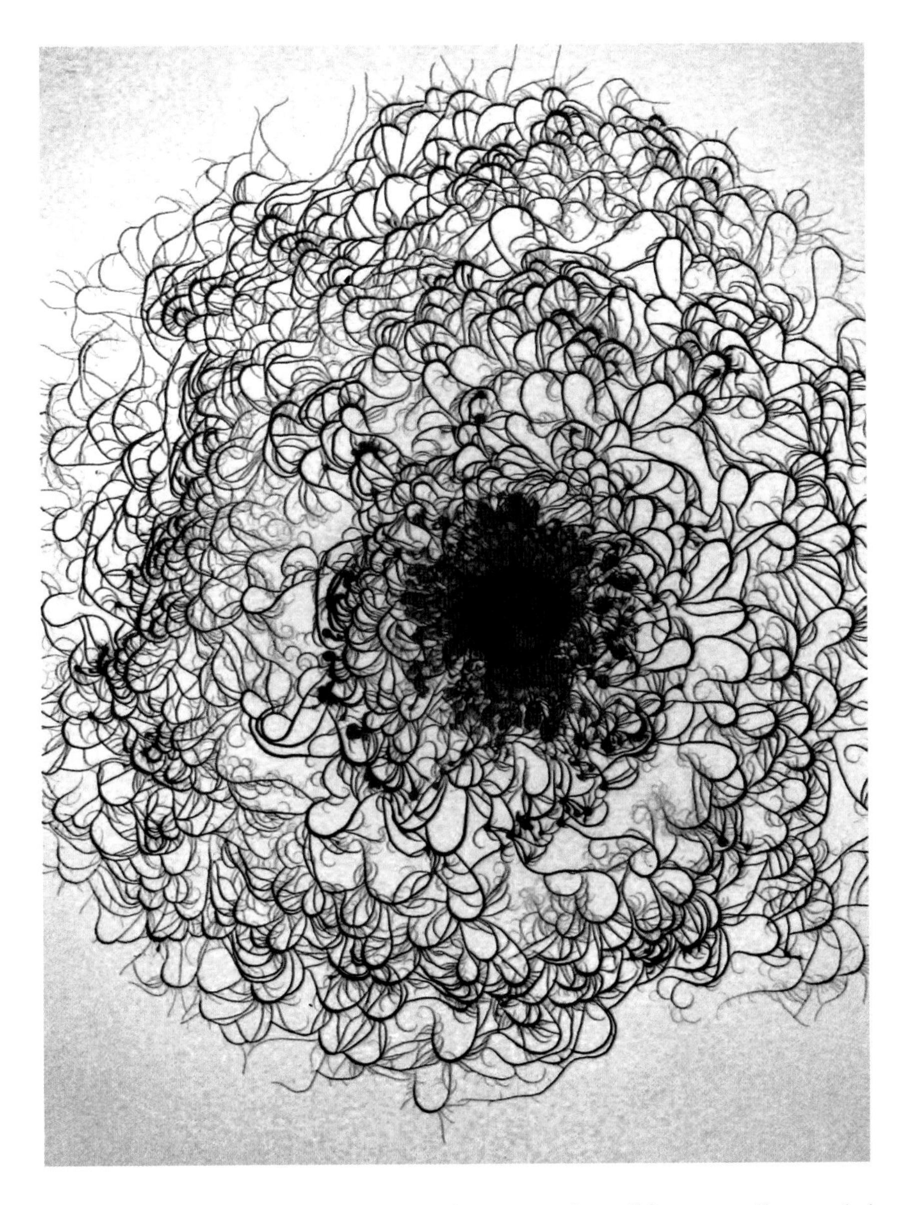

The patterns that bacteria produce depend on external conditions as well as on their intrinsic properties. These *Paenibacillus dendritiformis* of the chiral morphotype were grown under the same conditions as in the previous example but formed very different patterns.

By variation of the amount of food, moisture, and the firmness of their medium, the patterns formed by *Paenibacillus dendritiformis* keep changing. These cells grew on a hard surface (1.75% agar) with 5 g/L peptone. The picture shows part of a colony whose diameter is 7 cm.

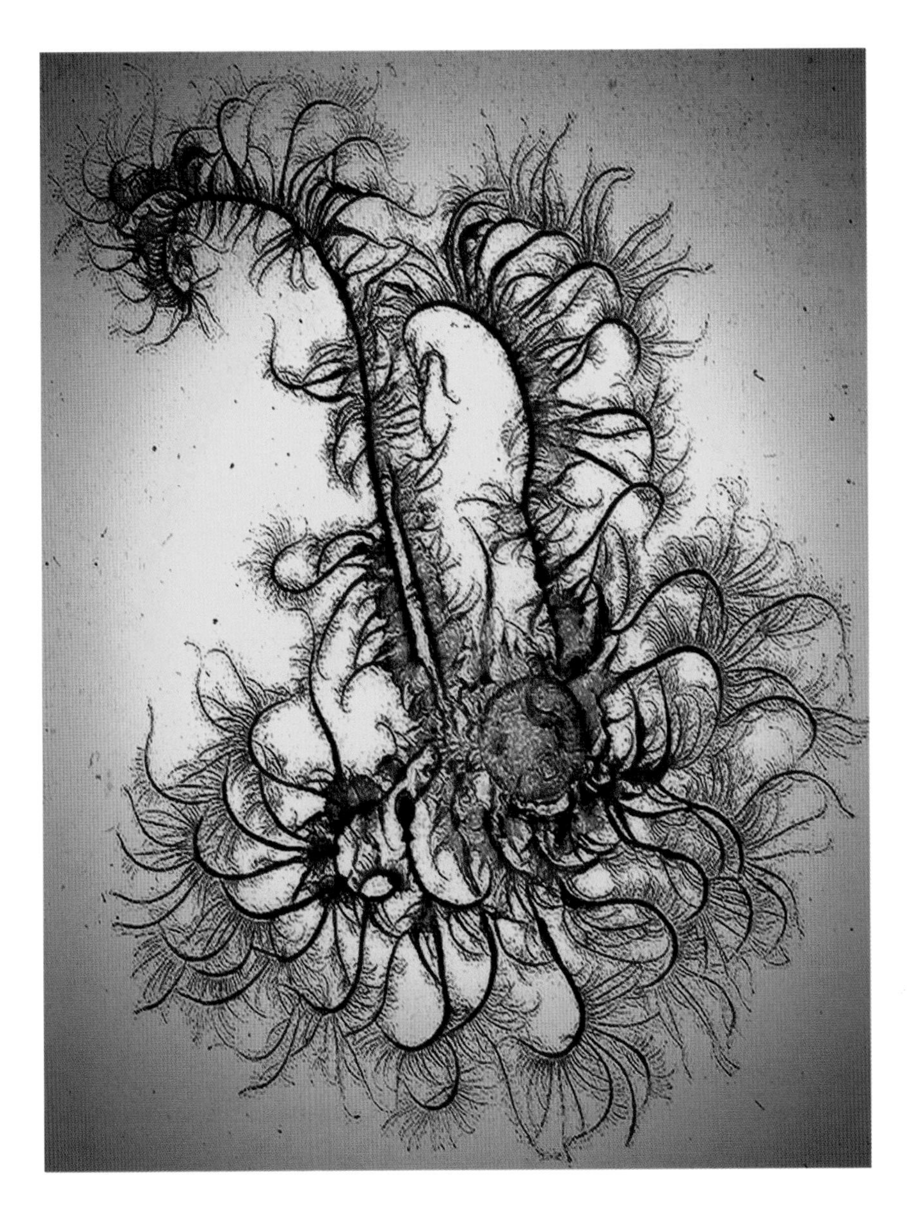

In this example of a *Paenibacillus dendritiformis* colony, the cells were grown on medium level of food (5 g/L peptone) and a hard surface (1.5% agar). The colony diameter is 5 cm.

This is a colony (width: 8 cm) of *Paenibacillus vortex*. The bacteria were grown with 5 g/L peptone on a very hard surface (2.25% agar). The bright dots are groups of tens of thousands of bacteria that move forward as one unit. These pave the way for the colony to expand on the very hard surface.
Source: This, and the previous color illustrations, are examples of bacterial art produced by Eshel Ben-Jacob.

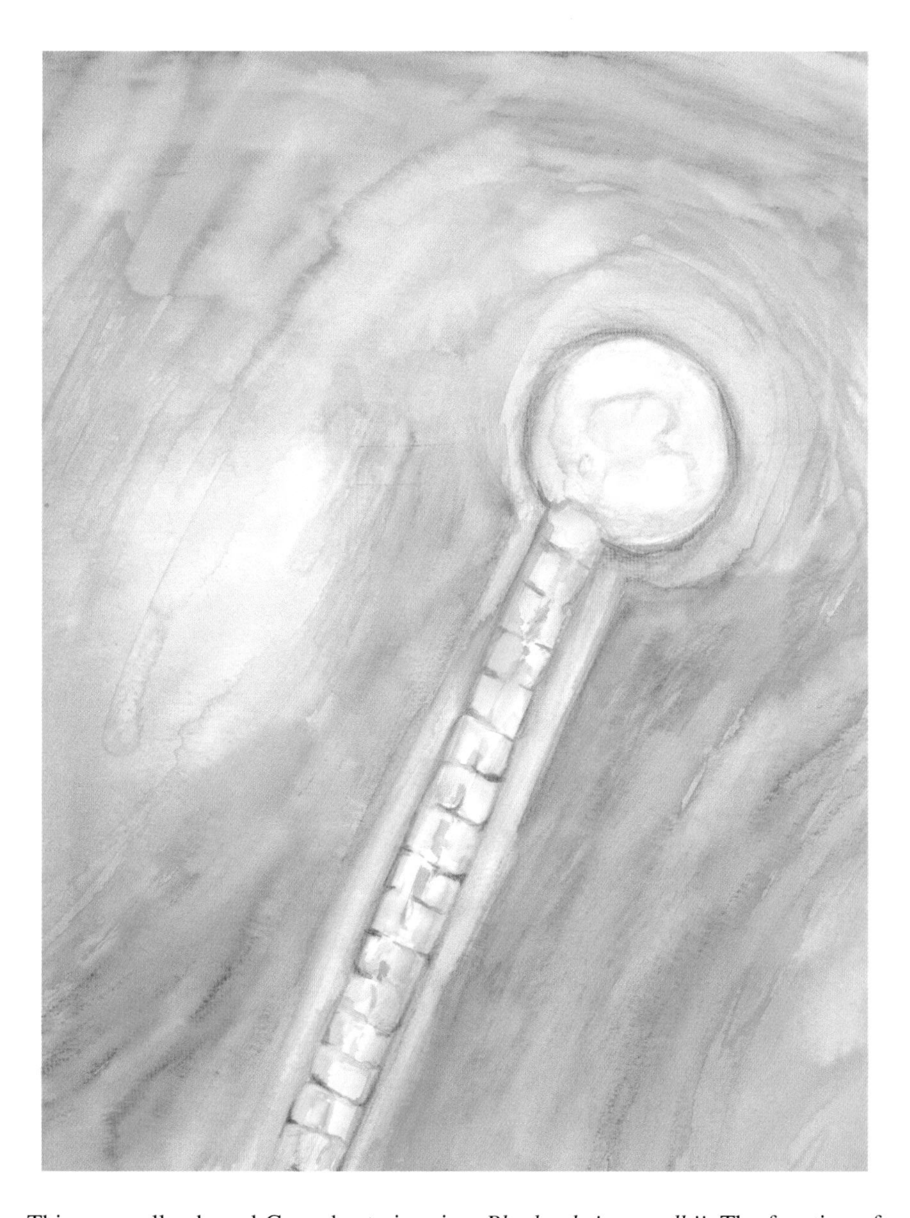

This unusually shaped Cyanobacterium is a *Planktothrix agardhii*. The function of the enlarged round cell at the end of this filament is not quite known.
Source: This water color and the next color illustrations were produced by Karoly Farkas.

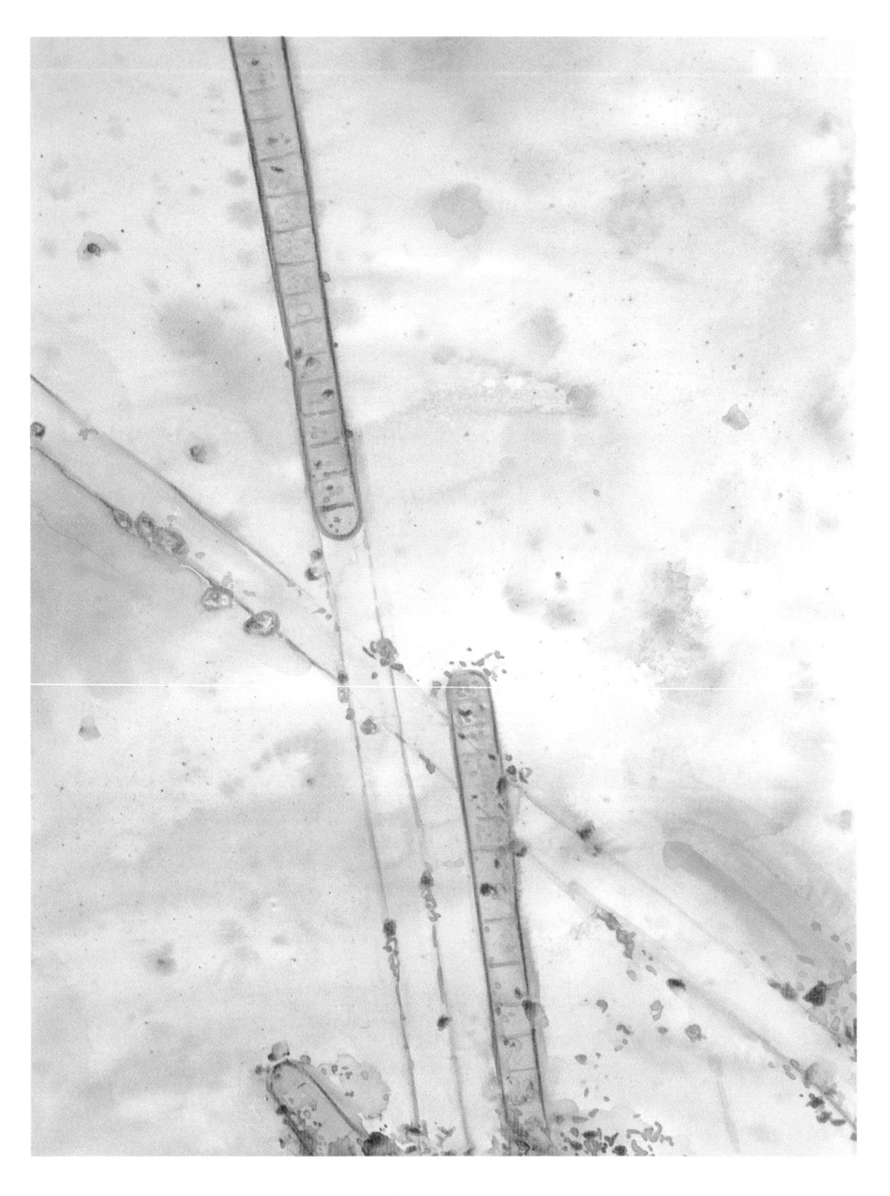

Filamentous Cyanobacteria look so fragile one can only admire their true powers. These cells belong to the genus *Lyngbya* but their species could not be identified.

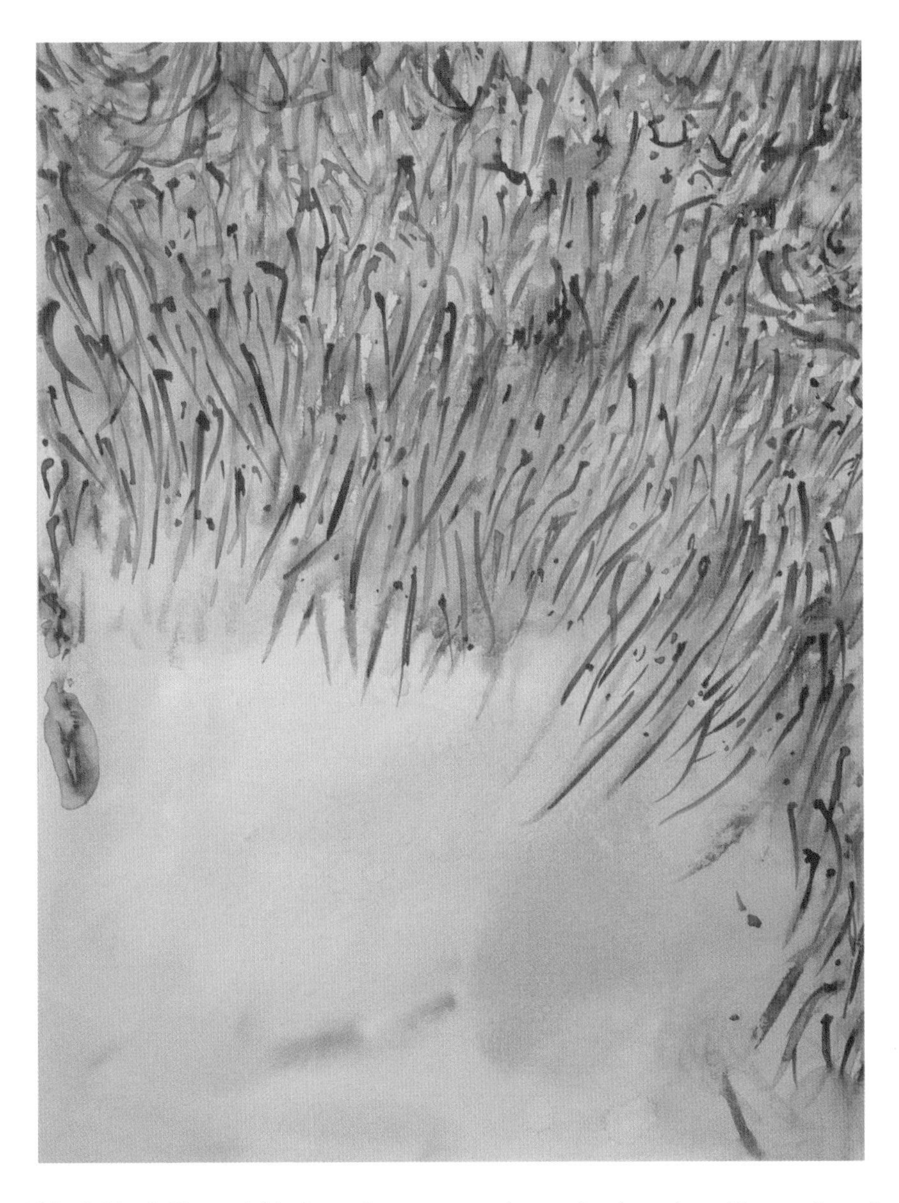

This field of *Gloeotrichia* bacteria appears to be partly cleared, as if a number of bacteria were wiped out. Indeed, these Cyanobacteria are common pray for grazing amoebas.

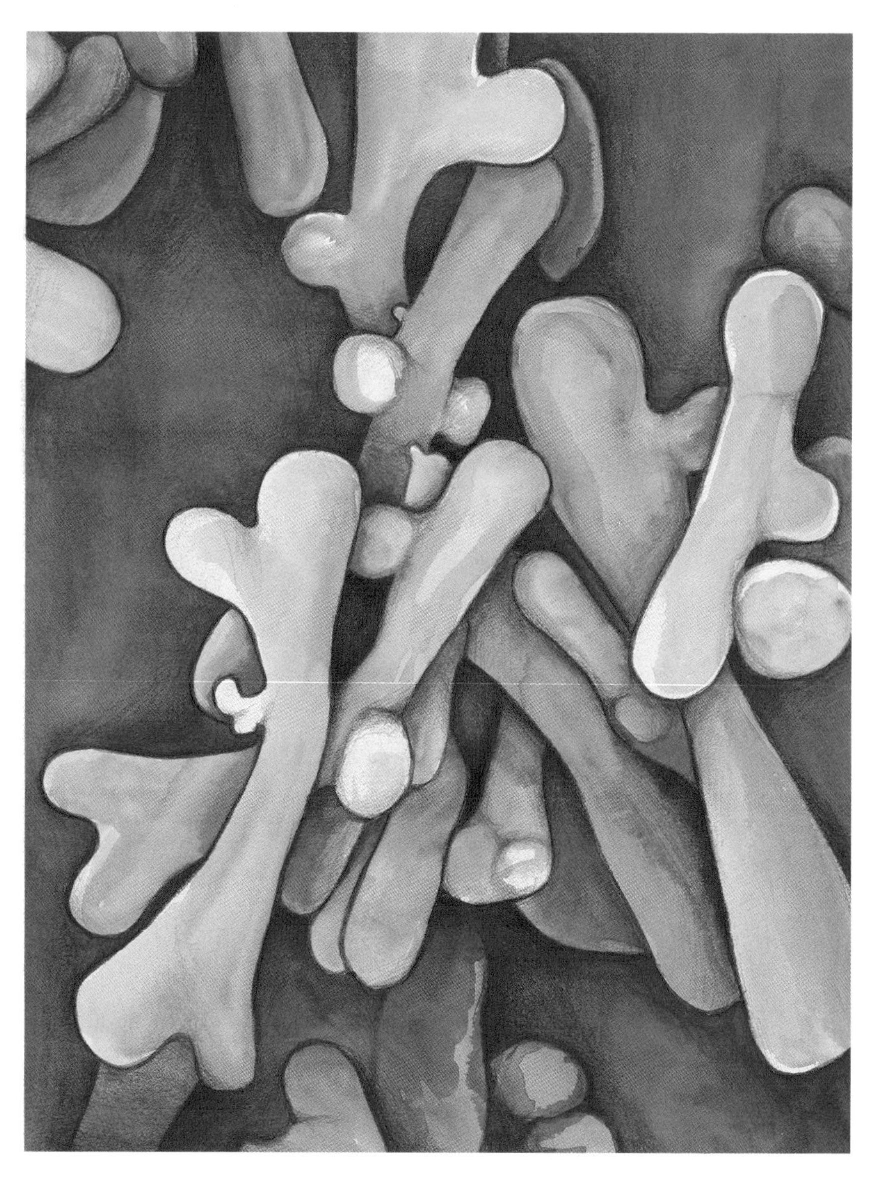

Bifidobacterium breve bacteria that are common in the infant gut have an unusual shape.

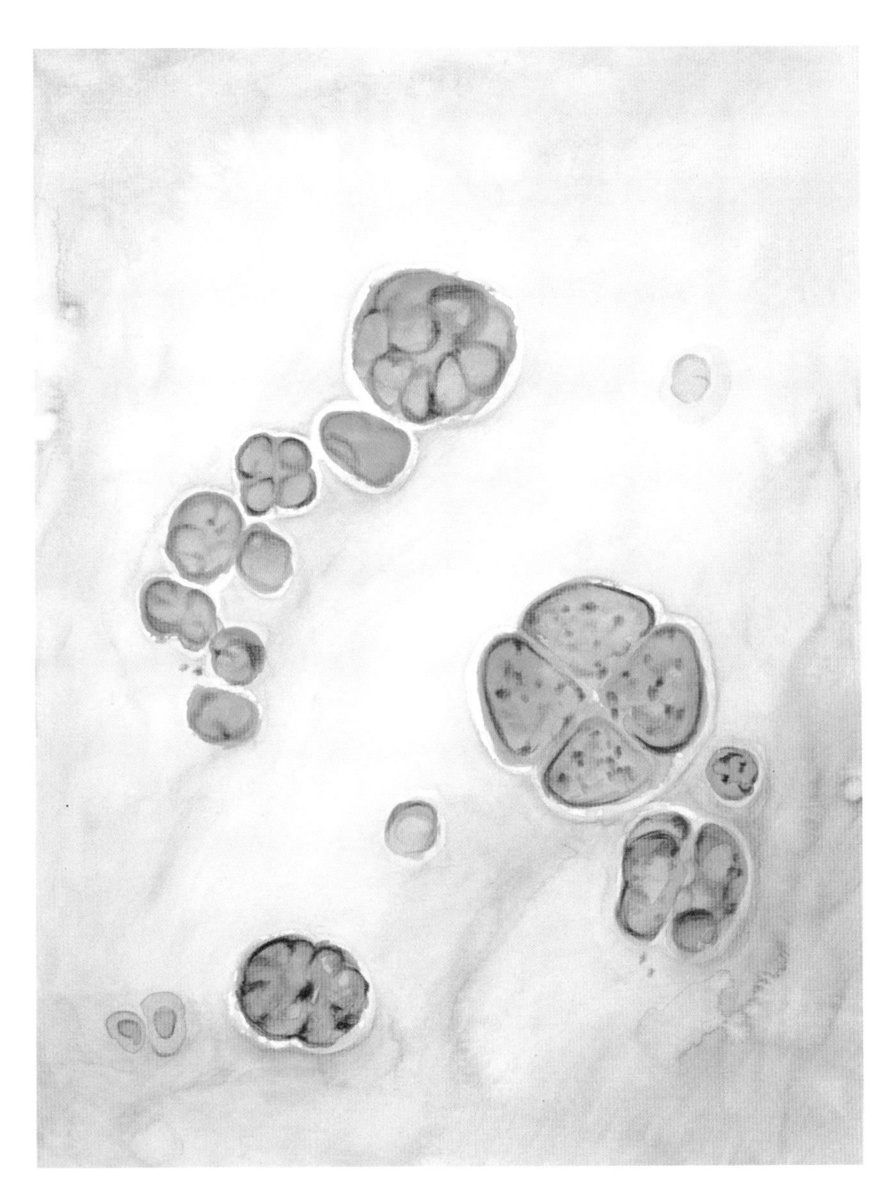

Cyanosarcina bacteria look like floating flowers. Relatively little is known about these Cyanobacteria.

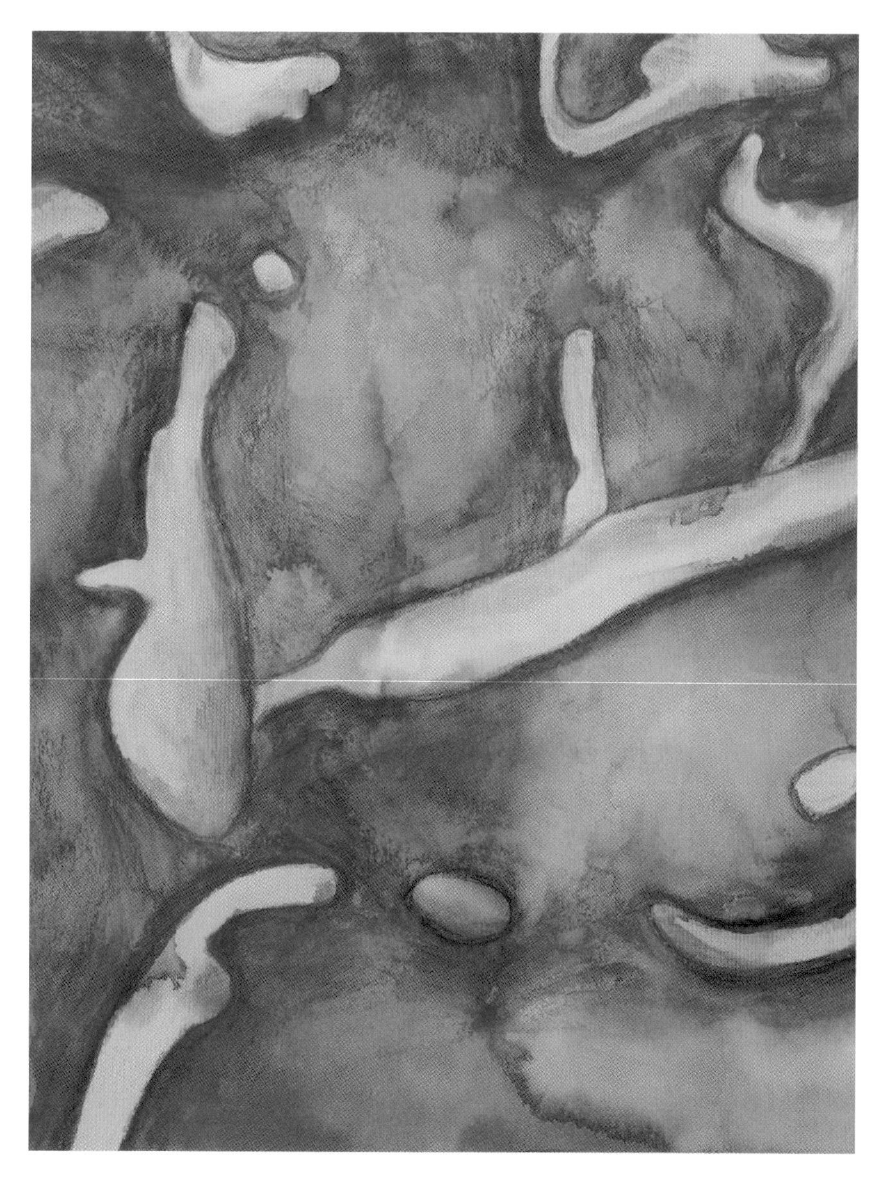

Bacterial shapes served as an inspiration source for this abstract work entitled "organic universe" *(detail)*.

Escherichia coli cells seem frozen in their busy lives, with their flagella entangled *(detail, private collection).*

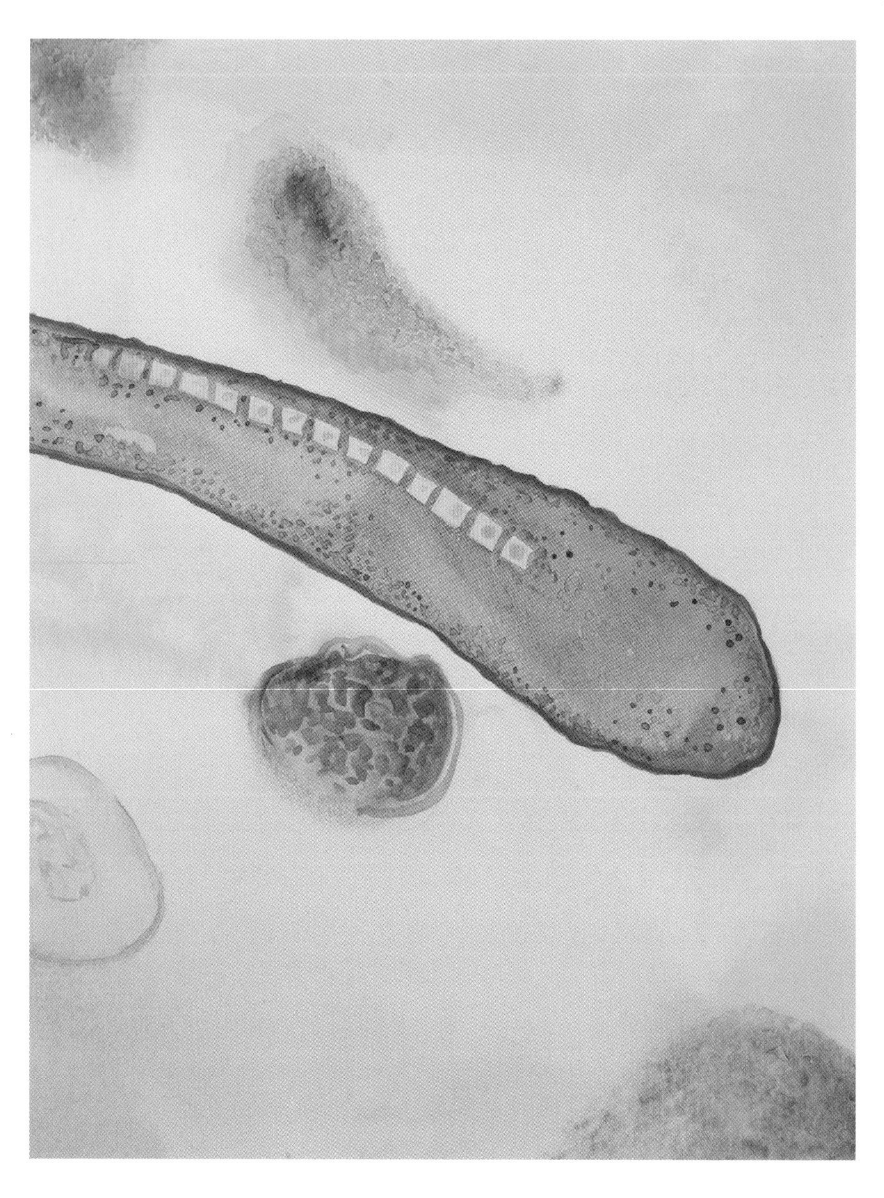

Magnetospirillum magnetotacticum belongs to the Proteobacteria. It can sense a magnetic field with its iron deposits (*detail*).

Tuberculosis has shortened the life of many artists or left its imprint on that of others, mostly writers. John Keats (England, 1795–1821) lost his mother to this disease, which is euphemistically known as consumption, a loss that made him study medicine. He soon gave up on this to dedicate his life to poetry, while his two brothers and a number of close friends all died of tuberculosis. When he got infected himself, probably from caring for his brother, he realized his life would be short, which may have contributed to the intensity of his works. He died at the age of 25, and he was one of many authors to succumb of tuberculosis. Jane Austen (England, 1775–1817) and later the sisters Brontë, who lived between 1816 and 1855, were tuberculous and frequently wrote about "fading women," "vanished bloom," and "wasted flesh," although their novels usually ended well, in contrast to many real-life stories.

Alexandre Dumas Sr. (France, 1802–1870) contracted cholera in 1832, but he survived, fortunately, or his works "Three Musketeers" and "The Count of Monte Cristo" would not have been written. His son Alexandre Dumas Jr. (France, 1824–1895) wrote "La Dame aux Camélias," based on a true story, which was the basis of the libretto that Guiseppe Verdi (Italy, 1830–1901) used for his opera La Traviata. It presents tuberculosis in the romantic view of those days, slowly killing a courtesan who is joined with her lover shortly before she dies. Forty years later, Thomas Mann (Germany, 1875–1955) wrote his book "Zauberberg" (Magic Mountain), featuring mostly characters suffering from this disease. He was inspired during his visits to his wife, who at that time was hospitalized because of tuberculosis.

Probably the most famous book that was written around a bacterial disease is the Decamarone, by Giovanni Boccaccio (Italy, 1313–1375). It is a medieval allegoric bundle of one hundred tales, supposedly told in 10 days by a group of 10 young people who had fled Florence to escape from the Black Death, also known as plague or the pest, the disease caused by *Yersinia pestis*. The stories describe in detail the physical, psychological, and social effects of that severe pandemic, which is further explored in Chapter 19. Early death due to infections was very common in the fourteenth and fifteenth century, and many pieces of art reminded people of their mortality. In France, paintings and frescoes appeared on walls of cemeteries, cloisters, and churches, which showed Death, depicted as a skeleton or an emaciated corpse, dancing with people of various social classes and professions. These "Danse Macabre" (Dance of Death) scenes may have started with a painting made in 1424 at the Cimetière des Innocents in Paris, which no longer exists, and became popular in France as well as in Germany and Italy.

Both frescoes and paintings depend on paint. Paint is a mixture of binders (oils, eggs, gums, or waxes), solvents, and surfactants, to which dyes or pigments are added to produce the desired color. Traditional pigments are inorganic metals (e.g., black oxide of iron, titanium white, or yellow ocher) and minerals (Egyptian blue) or organic molecules. The latter are mostly extracted from plants (like indigo, described in Chapter 15) and sometimes from animals (such as the deep-red pigment carminic acid, which is extracted from the scales of insects). The imperial purple that the Romans used to dye the toga's stripes of senators was isolated from *Murex* sea snails and was extremely valuable. More recently, such natural pigments have been replaced by chemically synthesized molecules. In future, bacteria may be a source of pigments. Many *Streptomyces* species (they belong to the Actinobacteria) produce pigments, which can be isolated and used to produce paint. *Streptomyces coelicolor* ("the color of heaven") produces, as its name suggests, a bright blue substance. *Rhodospirillum* bacteria, as we have seen in Chapter 13, produce red pigments, as do a number of salt-loving halophiles and other extremophiles (see Chapter 15). Although bacterial pigments are not yet being used on an industrial scale, they might assist in the production of biopaints in future. Scientists have already added the right genes to *E. coli* so that it produces indigo. The question as to why some bacteria naturally produce bright colors, however, has not yet been satisfyingly answered.

The fact that bacteria can be brightly colored means one can also use them directly to produce works of art. Already Alexander Fleming, the discoverer of penicillin, painted pictures by growing different kinds of pigmented bacteria on agar dishes, designed for esthetic reasons only. Nowadays, the repertoire of artists who want to use bacteria to create their works is widened with easy-to-grow bacteria in which the genes for various pigments have been genetically added. A variety of colors, including fluorescent ones, can be produced by genetically modified *E. coli*, which all grow at the same pace, so that a painting can be seeded in one instance; a full-color picture will appear after incubation as if by magic. This type of art is very short-lived, as the product will perish within days, but it is fun to experiment with bacteria in this way. As more people discovered this, web sites that are dedicated to microbial art were set up, two of which are mentioned in the Bibliography.

The binders that are essential ingredients of paints can be mistaken for food by bacteria. Fortunately, at least for the paintings, the metals lead, copper, and mercury, which are frequently present in pigments, are toxic to many bacteria, so they act as an antibacterial protectant. A study of the microbiological life thriving on sixteenth century paintings found that there were fewer species present when these metals were used in the paints. Many of the detected bacteria could not be cultured: only their telltale DNA could be demonstrated. The presence of bacterial populations means that they grow on these paintings—with the consequence that, over time, valuable components are degraded. When the paint starts to fall off the canvas, bacteria and fungi are a likely cause of the damage.

Not only oil paints but also frescoes are under attack by bacteria and fungi, although the pigments used lack binder components. Frescoes are paintings made in wet lime plaster using pigments mixed in water. The artist has to work quickly, coloring the picture section after section, while the plaster is still wet. Frescoes were popular as wall decorations in the Greek and Roman period and were again popular during the renaissance. The famous wall paintings on the interior of the Sistine Chapel in the Vatican (Rome, Italy) are frescoes made by Michelangelo Buonarotti (1475–1564). He was not specialized in the technique when he was asked (or nearly ordered) by Pope Julius II to decorate the chapel. Despite his lack of experience, he quickly learned and finished the work of a lifetime in four years.

Indoor frescoes may be protected against sunlight and weather, but they are not safe from bacteria and fungi, which add to the destructive forces of air pollution (including candle smoke), humidity, and temperature variation. A comparative study was performed to evaluate what microbes that were living on frescoes of medieval churches in Southern Italy could be cultured. The bacteria living on these fresco walls varied per church. The most numerous species that were found belonged to the Gram-positive *Bacillus* genus (Firmicutes); all samples contained *Bacillus* bacteria, although the species within this genus varied among churches. Also common, but found in lower numbers, were *Streptomyces* species and various fungi. Other studies have identified that *Nocardia* or *Micrococcus* species (also members of the Actinobacteria) and Gram-negative Proteobacteria (*Pseudomonas*, *Alcaligenes*) can also be found living on frescoes. Occasionally, the Bacteroidetes are represented by *Flavobacterium* species.

A lot of research on biological degradation of artwork, and how this can be prevented, is being done in Italy. That is understandable, as this country has the largest relative share of the world's cultural heritage. As odd as it may sound, the solution may be to add bacteria to threatened wall paintings: in particular, sulfate-reducing or nitrate-reducing bacteria are considered possible cures. These would minimize the amount of sulfur and nitrate that would then become limited for other bacteria that depend on these elements.

The oldest paintings made by man that we know of are rock paintings, which have been discovered in Africa, Australia, and Europe (to which this chapter is restricted). Our ancestors decorated rock surfaces, frequently in caves, with animal scenes, hand prints, or abstract patterns, for unknown reasons but with astonishing beauty. Of the Paleolithic rock paintings in France, the well-known Lascaux cave paintings are highly sophisticated. Discovered in 1940, it was subsequently determined that the paintings were over 13,000 years old. The caves were opened to the public in 1948 but had to be closed in 1963, because over 100,000 visitors per year caused too much damage to the site. Green algae had started to grow, photosynthesizing by the light of lamps and helped by the moisture introduced from human breathing. To prevent further damage, the surfaces were sprayed with the antibiotics streptomycin and penicillin to kill bacteria and with formaldehyde to deal with the algae. Despite these efforts, a few years later, thin white mycelium threads appeared, a sign of *Fusarium solani* growth (a mold), accompanied with *Pseudomonas* bacteria. Again antibiotics were applied. Nevertheless, black discoloring from various mold species stained the pictures a few years later. Bacteria were subsequently found to be abundant: especially *Ralstonia* and *Pseudomonas* species (both Gram-negative Proteobacteria). Sadly, paintings that survived for 13,000 years are now rapidly being destroyed by microbes, because people want to admire the ancient art. Caves are not museums, and we cannot regulate the microclimate well enough to conserve the paintings that had been protected in the sealed caves for so long. The solution for the Lascaux cave was to build a replica which visitors can admire instead of the original, and that is indeed the best way to protect such ancient art from destruction.

At the other extreme of our time scale, contemporary art frequently uses plastics, which can also be damaged by microorganisms. For example, the plastic ski cabin "Futuro", designed by the Finnish architect Matti Suuronnen in 1965, is still exhibited outdoors and represents an extraordinary document of its time. Its outside is composed of glass-fiber-reinforced polyester filled with polyester-polyurethane foam, and the windows are made of Perspex. These man-made surfaces are now heavily stained from bacterial growth; it was shown that both archaea and Cyanobacteria are to blame.

The degradation of our historical repertoire is not restricted to artistic works but extends to most historical information stored in museums and libraries. Paper, parchment, and papyrus are all subject to biodeterioration, unless this can be prevented with harsh remedies. Cynically, the oldest information storage medium, clay tablets, is not threatened. Subsequently used, much more vulnerable papyrus is made of the stem of the plant *Cyperus papyrus*. Strips cut from the long stems were aligned side-by-side to form two thin sheets that were wet mounted, one in horizontal and one in vertical orientation, after which they were pressed together under a weight and dried. Papyrus, which was usually produced as long scrolls, consists mainly of cellulose and lignin. Parchment came into use during the second century BC, presumably because the King of Egypt had forbidden the export of papyrus. Pergamon, in what is now Turkey, was known for its high quality parchment and gave it its name. It is a thin layer of animal skin and is mainly composed of collagen, a protein. In contrast to leather, which is tanned, parchment is prepared by drying the wet skin while stretching it.

The invention of paper is credited to a Chinese court official, Tsai Lun (also written as Ts'ai Louen), around the year 105 AD. In the eighth century, it was produced by Arabs in the Middle East, from where it spread to Europe. The Arabs introduced linen into the raw materials and built paper mills to scale up production. By the twelfth century, paper was the most common writing material in Europe. Paper consists mainly of cellulose, and as demand grew, wood became its raw material. The Mayans also wrote on a kind of paper, produced from the bark of fig (*Ficus*) trees. High quality paper required the addition of minor amounts of starch or gelatin to reduce the spread of ink, as well as dyes and pigments and, later still, whitening agents.

All these carriers of information and art are subject to microbial attack. Papyrus is mainly destroyed by fungi. When papyrus samples from museums in Cairo, Egypt, were examined, 38 different genera of fungi were detected, most of which were able to degrade cellulose. Papyrus can be completely decomposed by fungi or *Actinomycetes* (bacteria belonging to the Actinobacteria). Papyrus documents have to be protected against wasting away by the application of fungicides and antibiotics. Strict regulation of moisture is also required. Parchment can also be eaten by a number of fungi or bacteria. The latter were studied in detail on parchment documents from Russian libraries. Most of the isolated bacteria could multiply in the laboratory in a medium with parchment as the sole carbon and nitrogen source. Biocides (a combination of fungicides and antibiotics) are used to protect parchment. Paper's major threat are insects (whose gut bacteria take care of its digestion, as we have seen in Chapter 8), but fungi can also attack it, leaving strong pigment stains when they die.

The lamenting list can be extended to modern carriers of information and art. Textiles, man-made polymers, photographs, motion picture films, and even magnetic tapes or optical CDs are all subject to biological degradation. Photographs are mainly attacked by fungi but also by *Streptomyces*, which appreciate the presence of substances of animal origin, such as gelatin or albumin. Magnetic media are not only sensitive to heat, light, grit, moisture, and atmospheric pollutants but also to microbial threats such as fungi and, again, *Streptomyces*. It seems that, as time progresses, our inventions of information carriers become ever shorter lived. The shelf life of photos and magnetic tapes is shorter than that of paper, and recent paper products are less resistant to degradation than early paper, which loses again, in terms of long-lasting duration, to parchment and papyrus. The safest way to store valuable information for centuries and millenniums are still clay tablets. Those are indigestible even for bacteria.

18

Fixing the Air

The air contains three gases that are in flux with the biosphere: nitrogen gas (N_2), oxygen gas (O_2), and carbon dioxide (CO_2). Air consists for the most part of nitrogen: approximately 78% of air is nitrogen gas, whereas only 21% is oxygen and carbon dioxide exists only in minute amounts, as was presented in Chapter 14. Because of its abundance in the air, this chapter is dedicated to the processes by which nitrogen gas is taken up from the air to be incorporated into biomatter and how it can subsequently be released back into the atmosphere. Bacteria are responsible for all these processes.

Farmers have known for a long time that particular plants of the Legume family can "improve" the soil. Legumes, to which beans, mimosa, and clover belong, are particularly known for this magic, but they are not solely responsible: the plants need bacteria to put biologically active nitrogen into the soil, which is the basis of this natural fertilization. Most organisms cannot use nitrogen gas as a source of nitrogen, an element that is essential for all life forms. For most living organisms, nitrate ions (NO_3^-) are a good source of the element nitrogen, which they need for their proteins, DNA, and RNA. Some organisms prefer ammonia (NH_3) or nitrite ions (NO_2^-). The exceptions are *nitrogen-fixing* bacteria, which can transform nitrogen gas into biomatter.

Bacteria that can fix nitrogen are nearly all Gram negative and belong to the phyla Cyanobacteria or Proteobacteria. *Azotobacter* species (a genus of the latter phylum) and *Cyanobacterium* species (which is the genus that named this phylum) can fix nitrogen all by themselves. *Rhizobium* species (another Proteobacterium),

Bacteria: The Benign, the Bad, and the Beautiful, First Edition. Trudy M. Wassenaar.

on the other hand, live free in the soil, but in order to perform their nitrogen-fixation skills, they need to grow inside special nodules that plants of the Legume family provide. These plants can also live without their bacterial friends, but for both *Rhizobium* and Legumes to live independently, the soil must contain nitrate (or ammonia, which is quickly transformed to nitrate by other soil bacteria). Their partnership is an example of mutualism, where bacterium and plant help each other (in this case to transform nitrogen gas into ammonia food), but they can also live independently, which sets it apart from true symbiosis. Nitrogen-fixing bacteria of the genera *Frankia* (an exceptional member, in this chapter, of the phylum Actinobacteria) and *Azospirillum* (Proteobacteria) live in mutual relationship with plants other than Legumes.

The mutualism of Legumes and *Rhizobium* works as follows. The roots of the plants are first "invaded" by the bacteria present in the soil, after which the plant reacts by producing a specialized chamber for the bacteria, the root nodules. These are visible as little round knobs on the roots of the plant. Every Legume species hosts its own kind of *Rhizobium*. This specificity is mostly dictated by the bacteria, which are selective by means of their TTSS (see Chapter 5). They secrete effector proteins inside the root cells of the desired plant species only, after which these welcome them by offering them a home.

The plant provides the bacteria with high energy sugars, which they convert to ATP. A lot of energy, and thus ATP, is needed for nitrogen fixation. The gas is relatively inert, which means it has to be persuaded to react at all; the chemical way to force reactions that do not take place spontaneously is to add energy. In the bacterial cell, enzymes force such reactions with the use of ATP. The crucial enzyme for N_2 fixation is nitrogenase. In the case of Legume-*Rhizobium* mutualism, this enzyme is provided by the plant. Obviously, *Azotobacter* produces its own nitrogenase, as otherwise it could not fix nitrogen independently.

Nitrogenase is an enzyme that requires one atom of molybdenum and a number of iron atoms (both are metals) in order to function. There are other enzymes that use metals such as iron or zinc, but the requirement for molybdenum is unusual. This element happens to be rare in eastern Australian soils, where its limited availability contributes to a relatively infertile soil. By adding only small amounts of molybdenum, growth of clover can be stimulated, which, with the help of *Rhizobium*, can fix nitrogen. And once the clover plants have fixed nitrogen and integrated it into their biomass, it is, and remains, available for other organisms. Molybdenum-dependent clover culture has been used on a large scale to fertilize the soil, after which agricultural exploitation became feasible. Another feature of nitrogenase is that it becomes inactive in the presence of oxygen. This is why the *Rhizobium* bacteria live in the plant's root nodules: these chambers are kept oxygen free.

Azotobacter species live freely in soil and are able to use nitrogen without the help of plants. It is relatively simple to isolate *Azotobacter* bacteria, just by cultivating a bit of soil in a medium that does not contain any nitrogen source. Nothing will be able to grow, other than *Azotobacter*, which are now forced to perform their nitrogen-fixation chores. They do so even in the presence of oxygen, although their nitrogenase is just as sensitive to oxygen as that of other bacteria. *Azotobacter* solves this problem with a very high degree of respiration, which keeps the intracellular oxygen levels low enough for nitrogenase to remain active.

Besides some Proteobacteria, to which both *Azotobacter* and *Rhizobium* belong, the Cyanobacteria are mostly known for their capacity to fix nitrogen. These are main players (together with archaea) for nitrogen fixation in the ocean. For a long time, scientists thought that the Atlantic was the main place where oceanic nitrogen fixation took place. They realized that iron was a limiting factor for Cyanobacteria to grow, and because the Atlantic receives iron via frequent winds that bring dust from the African continent, it was thought most Cyanobacteria would live there. Now it is suggested that, in fact, the Pacific Ocean, which is far bigger, hosts most of the marine nitrogen fixers. If iron is the limiting factor, then obviously the Pacific gets enough iron from whatever source to host large numbers of Cyanobacteria.

Not all Cyanobacteria can fix nitrogen, and not all live freely in the ocean. Some prefer a terrestrial life. A variety of these species live as symbionts to plants, including ferns or grasses, and some are symbionts to fungi. The Cyanobacteria are sometimes described as simple algae or even as plants because they can fix carbon dioxide, as we have seen in Chapter 13. But Cyanobacteria lack a nucleus, mitochondria, or chloroplasts, and the RNA of their ribosomes clearly identifies them as bacteria. They neither are algae nor are all blue-green, despite their old description as "blue-green algae." Various species have different colors, and some may have given the Red Sea its name.

Cyanobacteria are little wonders of Nature. Many of them can literally live off air, as they can use both nitrogen gas as a nitrogen source and carbon dioxide as a carbon source. All they need is light for their photosynthesis, minute amounts of other elements such as metals, and a bit of phosphor, preferably in the form of phosphate, plus even smaller amounts of sulfur. (The nutrients that mostly limit Cyanobacterial growth in the ocean are probably phosphate and iron.) No wonder these bacteria were proposed as the first inhabitants of our planet, as they seem pretty much independent of other life forms, which few other organisms can claim. When the volcano on Krakatoa (an island between Java and Sumatra, in Indonesia) erupted in 1883, the original vegetation was exterminated by the blast. Legend has it that Cyanobacteria were the first organisms to recolonize the bare island, although in those days, methods to determine bacterial growth were limited. It can no longer be checked whether indeed Cyanobacteria were the first to arrive, but in view of their independent lifestyle, the story could be true.

Cyanobacteria even live a little like multicellular organisms. The filamentous Cyanobacteria, to which *Anabaena* species belong, develop chains of cells that have different tasks, depending on demand. When no other nitrogen source is available, a few cells along the typical multicellular filament specialize to fix nitrogen. These are interspersed between photosynthetic cells living off that nitrogen. The photosynthesizers are in the majority, as they provide the energy (ATP, which they produce from sunlight) that the nitrogen-fixing cells need. The few nitrogen-fixing cells are slightly larger than the more abundant photosynthetic ones, so that the filament looks a bit like a rosary. This is yet another way of solving the incompatibility between photosynthesis, which produces oxygen, and nitrogenase, which cannot stand the stuff. (An alternative strategy, followed by other kinds of Cyanobacteria, is to perform photosynthesis during the day and fix nitrogen at night.) But that is not all. When life gets hard, especially when phosphate supplies become limited, *Anabaena* can form sporelike cells that wait for better conditions. Lastly, they can develop into a motile bunch of gliding bacteria (see Chapter 4). Thus, four different cellular forms within the same species exist, and, depending on the need, cells develop into one or the other. The border between unicellular and multicellular organisms is blurred by these and other life forms, such as the social Myxobacteria, mentioned in Chapter 4. No matter how much scientists like labels, we have to live with the fact that Nature may not always divide things into black and white but can offer many shades of gray.

It could nearly be forgotten that eubacteria are not the only prokaryotes to fix nitrogen. Some archaea can do it, too, and they use a similar molybdenum-dependent nitrogenase. So far, it has been described only for members of the phylum Euryarchaeota, and more specifically for the methanogens within this branch. These are archaea that produce methane and live a life without oxygen. Methanogens live in the intestines of ruminants, where they are responsible for large amounts of methane production, and in rice fields. Ruminants and rice fields are two major biological sources of atmospheric methane, which is a far more potent greenhouse gas than carbon dioxide. Methanogens will turn on nitrogen fixation when external nitrogen sources are limited. Nitrogen-fixing methanogens also live in the ocean. These anaerobic archaea might even have been the inventors of nitrogenase, in which case Cyanobacteria would have learned this trait from these archaea, but currently that is still speculation.

Once nitrogen has been incorporated into biomatter, most of it will be recycled. The nitrogen present in living or dead organic matter, where it is found in biomolecules (proteins and nucleic acids), eventually degrades to nitrate or ammonia, depending on which bacteria, fungi, or animals feed on it. Animals may produce urea as a waste product (the main component of urine), which degrades into ammonia spontaneously. Some bacteria, notably *Nitrosomonas*, can transform ammonia into nitrite, and this nitrite is converted to nitrate by others, for instance, *Nitrobacter* species. Both add oxygen to a nitrogen-containing molecule, which releases energy (just like the burning, or oxygenation, of sugar does). In fact, these organisms can fix carbon (which was treated in Chapter 13) without the need for light, as they get their energy from the chemical oxidation of these nitrogen compounds. They are called *chemoautotrophs*, where the "chemo" identifies chemical reactions as their energy source and "auto" stands for being self-supportive in terms of carbon: they do not need an external carbon source. These processes are very efficient in soil, where most nitrogen, when not incorporated into biomatter, will end up as nitrate sooner or later. Nitrate is a form of nitrogen that can be used by plants and animals alike, so all this nitrogen can cycle within the biosphere. Eventually, substantial amounts of terrestrial nitrate leak out of the soils into effluents, and rivers transport it into the sea, where it can be used by marine life.

But who puts nitrogen back into the atmosphere, to compensate for the nitrogen that nitrogen fixating organisms have dumped into the biosphere? A geological source for N_2 is volcanoes, which constantly puff out the inert gas, but the prokaryotic world partly takes care of that, too. There exist both eubacteria and archaea that metabolize nitrogen from biomatter and produce nitrogen gas as a waste. This is called *denitrification*, as it removes nitrogen from the biosphere, through the intermediates nitrite and nitrate. Both terrestrial and marine bacteria can do it, but only in the absence of oxygen. To transform oxidized nitrogen compounds (nitrate, nitrite, or nitrous oxide) to nitrogen requires electrons, and bacteria would prefer to donate these to oxygen, if present. The bacteria that can produce N_2 as a waste product do so because they need an alternative electron acceptor when oxygen is absent. The need for electron acceptors results from burning sugars (respiration). When sugars are burned without oxygen, it is called *anaerobic respiration*. Denitrifying bacteria can live either with or without oxygen, and this condition determines whether N_2 will be produced or not. Typical environments that are anoxic are the deep soil and stagnant water or swamps, and the workers here may be *Pseudomonas aeruginosa*, *Thiobacillus denitrificans*, or *Paracoccus denitrificans* (all Gram-negative Proteobacteria), as well as some *Bacillus* species (Gram-positive Firmicutes). The enzyme responsible for the last step, which makes nitrogen gas from nitrous oxide, is called nitrous oxide reductase; it contains copper (all enzymes involved in the process need metal ions). Denitrifying bacteria also live in the gut of earthworms, which as a consequence poop nitrogen gas.

Denitrifying bacteria were first isolated in pure culture in 1886, a few years before their counterparts, the nitrogen-fixing bacteria, were discovered. Ulysse Gayon (1845–1929), together with Gabriel Dupetit, from France, managed to isolate what they called "*Bacterium denitrificans*". Unfortunately, the culture no longer exists, so we cannot be sure what species they had isolated. Gayon specialized in the quality of wine later in life, as he lived in Bordeaux. Denitrifying bacteria help to remove nitrogen from sewage and municipal wastewaters. Despite this useful activity, they used to have a bad reputation, as their activity depletes the soil of nitrate, which makes it less fertile. Fertilizers, either of natural (manure) or of chemical origin, are therefore used to add nitrate back to the soil (the other main components of fertilizer are phosphorus and potassium). In some locations, fertilization is overdone a bit. The concentration of nitrate in ground water has locally increased to unacceptable levels, and too much nitrate in drinking water is unhealthy. Nitrate is one of the most troublesome pollutants of surface water, and denitrifying bacteria are now perceived as beneficial to get the problem under control.

Although this story was taught as a complete picture to generations of students, some facts remained unpleasantly unexplained. Notably, it seemed that far more nitrogen disappeared from the oceans and marine environments than could be accounted for. The riddle was solved with the recent discovery of the Anamox bacteria. Anamox stands for anaerobic ammonium oxidation: the conversion of ammonium, via nitrite and nitrate, to N_2, all in one cell, in the absolute absence of oxygen. It is performed by bacteria of the phylum Planctomycetes. Most Anamox bacteria live their slow life (they take their time multiplying, at least in the laboratory) in the ocean or in estuaries, marshes, rivers, and lakes, as long as oxygen can be avoided. An astonishing feature in their metabolism, which has not been completely resolved yet, is that one of the nitrogen intermediates they produce is hydrazine, the explosive fuel that rockets use. Although slow growing, these bacteria are extremely efficient at removing nitrogen from their surroundings and returning it to the atmosphere. Since their discovery in 1995, it has been calculated that Anamox bacteria are responsible for over 50% of all marine nitrogen that is put back into the atmosphere. This, along with the discovery that the ubiquitous archaea in the ocean can perform denitrification, has revolutionized our view on the global nitrogen cycle. It shows that novel discoveries in microbiology can still be made, and these can have huge implications on our insights of processes taking place on a global scale.

19

Pest and Pestilence

Bacterial diseases should not fail in a book that promises to describe "bad bacteria," so here are three diseases that stick out when describing major epidemics or threats: anthrax, plague, and cholera. These are very different diseases. Anthrax is caused by a Gram-positive species that produces spores (see Chapter 3); its virulence depends on a toxin that meddles with the host cell's cyclic AMP levels while a second toxin bores holes in the membranes of host cells, as explained in Chapter 9. In addition, the cells of *Bacillus anthracis* have an armorlike polymer coat (a "capsule"), which protects them against the host's immune system. Plague, on the other hand, is the result of *Yersinia pestis*, a Gram-negative Proteobacterium that requires an insect vector to spread, is dependent on a TTSS (see Chapter 5), and enters the host's cells to wreak havoc: it is "invasive". *Vibrio cholerae*, another Gram-negative Proteobacterium, causes massive diarrhea by means of its toxin, but it mostly lives in the sea, as we will see. Cholera still causes large outbreaks in various parts of the world. In contrast, anthrax and plague may not be the major worldwide killers as their negative reputation foretells, but fear for these diseases can still cause social upheaval and mass panic outbreaks. They deserve to be treated with respect, and this chapter is reserved for them.

Anthrax is mostly a disease of animals, but humans can catch it by inhalation or ingestion of the spores formed by *B. anthracis*. Inhalation of sufficient numbers of anthrax spores causes a lung infection for which the medical term pulmonary anthrax is used. Once the spores enter deep into the lungs, they encounter an environment that is paradisal, at least for a spore: humid, warm, and nutritious.

Bacteria: The Benign, the Bad, and the Beautiful, First Edition. Trudy M. Wassenaar.

The spores come back to life, and the bacteria start to multiply. Without immediate and effective antibiotic treatment (macrolides are typically used), the infection can kill a patient within days. Scary as this sounds, one would need to inhale high numbers of spores in order to get infected.

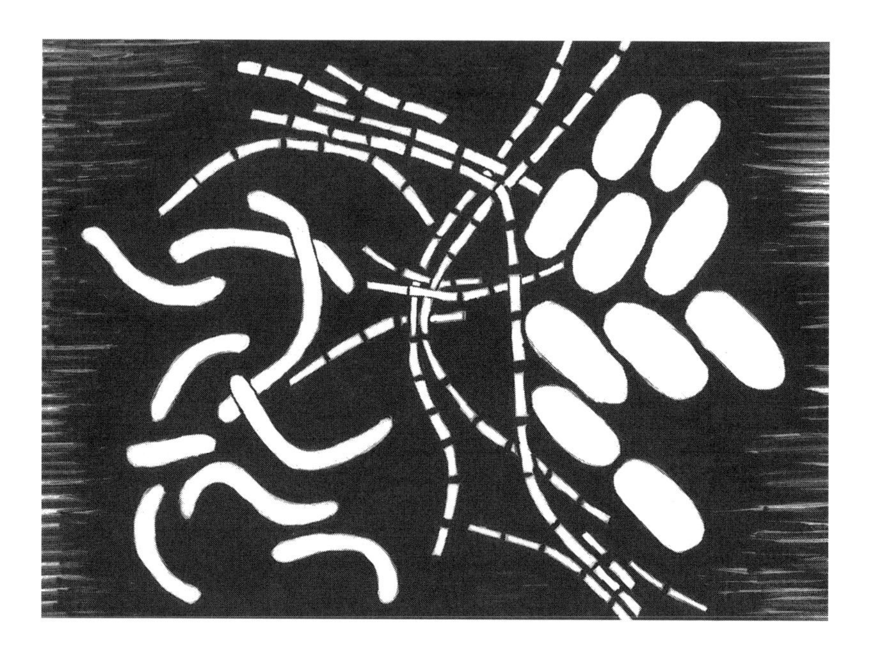

Ingestion of the spores, through eating contaminated food (for instance, by consuming infected game), can cause infection of the intestine, which is just as dangerous as a lung infection. A third way in which anthrax bacteria can enter a body (which they have to do in order to multiply) is directly via a wound, but these infections are less frequently lethal. The natural manner in which anthrax bacteria multiply, independently of humans, is intestinal infection of grazing animals that ingest spores present in the soil. The poor animal will die as a result of anthrax bacteria multiplying in its blood. The cadaver will eventually disintegrate, and when *B. anthracis* is exposed to air, it will turn into spores. These contaminate the soil, and the circle is completed: the spores will be waiting for a new victim, no matter how long it takes.

Anthrax is one of the few bacterial diseases that are spread from dead bodies. Although it is generally perceived that dead bodies and cadavers are dangerous, most infectious diseases are in fact spread by living animals and people. Anthrax is an exception, as the deadly spores are mostly liberated from decomposing infected bodies. The disease was dreaded already in historical times. The fifth plague that the Bible describes in the book *Exodus* could have been anthrax (although other diseases would fit the rather vague description, too), as well as the "burning plague" described in Homer's Illiad. Virgil (70–19 BC) described in detail an epidemic of what must have been anthrax and noted that the disease could spread from animals to humans. In the Middle Ages, it was known that animals that had died of anthrax were dangerous to handle. Some pastures were "doomed," as animals grazing there would fall ill and die. Burial of the cadavers was not sufficient to break the contamination chain, as the spores could still spread in the soil. The latter was not known, of course, but people learned it was safer to either burn the cadaver, bury it with quicklime (calcium oxide) to minimize the risk of spread, or let bushes grow over the burial to prevent grazing at that location.

Anthrax was also known as woolsorter disease, since wool sorters, who had to handle wool from healthy and diseased sheep alike, were frequently exposed to the disease. The workers could get infected from inhalation of anthrax spores, which only few survived. It became particularly common in the nineteenth century in the Bradford area of the United Kingdom, where the wool industry was blooming. The disease dramatically increased in frequency with the import of mohair (hair from the Angora goat imported from Turkey) and alpaca (llama hair from Peru), and soon the disease was common, in particular for sorters handling these types of wool. Once the cause of the disease was understood, its transmission to humans and animals could be minimized, and anthrax became a rare disease.

The interest in (and fear of) anthrax exploded in 2001, when postal delivery in the United States brought anthrax spores indoors. Five letters, addressed to news organizations and two US senators, were posted only one week after the 9/11 terrorist attacks that shook the country. These letters contained a white powder in which, as was found out two weeks later when the first victim fell ill, large amounts of anthrax spores were present. Twenty-two persons got infected, of whom five died. Congressional offices and mail-processing centers had to be decontaminated, while every unidentified white powder caused panic. The hunt for the culprit took seven years and cost $60 billion; the cost in terms of threatening mass panic and loss of trust was immeasurable.

The presumed perpetrator committed suicide in 2008, just before he had to face charges. Bruce E. Ivens was a scientist at the US Army Medical Research Institute of Infectious Diseases (USAMRIID), where a particular strain of *B. anthracis*, called the Ames strain, had been a subject of research. Since Ivens had worked on an anthrax vaccine program, he knew the ins and outs of spore production, and he had been immunized against the disease (an important factor, as an unvaccinated individual would not have survived the handling of these letters). His research program had come under attack and was threatened with termination. The spores present in the letters contained a genetic footprint that could be traced to his lab, which in turn eventually lead to the suspect, but only after the genomes of multiple isolates had been completely sequenced.

The investigators had a hard time solving this case. Classic forensic research was hampered by the fact that all objects of evidence were contaminated and needed to be handled under Biosafety Level Three precautions. The researchers who were able to solve the genomics puzzle were also potential suspects, when it became clear that the strain used in the letters was a frequent subject of research. It meant that from a team of researchers, each individual was given only part of the puzzle to solve, without knowing the big picture, in order to avoid manipulation. In retrospect, it became obvious that Ivens at one point had indeed tried to hamper investigations. He had not reckoned with the minute mutations that accumulate when bacteria multiply, which produced the telltale genetic footprint that eventually solved the case.

As dramatic as this story is, a historic outbreak puts the toll of this disaster into perspective. On June 3, 1770, Port-au-Prince, the capital of Haiti, was hit by an earthquake that devastated the city and its surroundings. The local society collapsed, and famine soon struck the thousands of slaves living on the island, especially in the Saint-Domingue area. In an attempt to combat hunger, a large amount of salted and smoked, but uncooked, beef was bought from Spanish traders—who were glad to be rid of the meat, as it had originated from sickened cattle. Within six weeks, over 15,000 people died from eating what turned out to be anthrax-contaminated meat. This epidemic is not generally known, as it never received much attention, but it illustrates how deadly anthrax can be and how important surveillance of the food supply is.

It is not always easy to identify the causative agent of historical epidemics. Even for the Haitian outbreak of 1770, some scientists propose that bubonic plague, rather than anthrax, caused the epidemic. Inaccurate accounts of symptoms and confusing terminology (the terms "pest" or "plague" were used as general terms to describe epidemic diseases) are at the bottom of this uncertainty. Even the most famous and dreaded "Black Death" epidemic of fourteenth century Europe is proposed by some to have been caused by something else than plague, although that still is the most likely cause and fits most descriptions.

The disease plague, which occurs in three different clinical manifestations, is caused by *Y. pestis*. The organism can survive and multiply only by frequent exchange between a mammal, usually a rat, and its fleas. The bacteria multiply in the flea's intestine, and when numbers are high enough, they block the flea's sucking organ, which leaves the insect permanently hungry, so that it will frequently bite the hosting animal. With every bite, a few bacteria are secreted into the bloodstream, where they can multiply. Such an infected animal (which will eventually die of the disease) is a source for other fleas to contract the bacteria. Infected fleas that move on to the next animal spread the disease.

People can become infected through two different routes. When infected fleas (which do not experience symptoms) leave their rodent host to bite a human, they inject bacteria into the bloodstream. These travel to the lymph nodes, where they start multiplying. The lymph nodes enlarge, forming a "bubo," a local swelling that is typical for bubonic plague (one of the three clinical manifestations). In unlucky individuals, high numbers of bacteria leak into the bloodstream to give septicemic plague, the second clinical outcome. Bacteria growing in the blood shed large amounts of membrane debris, called LPS (short for lipopolysaccharide, a mixture of lipids and sugar polymers), which sets the immune system in overdrive. Most patients die at this stage from septic shock, a direct result of the LPS in their blood. An alternative, third outcome of plague is when bacteria reach the lungs, to give pneumonic plague. Now, coughing patients can contaminate other persons through aerosols, the second route of transmission. A person who got contaminated from such aerosols will immediately develop pneumonic plague, which progresses much more rapidly and is more deadly than bubonic plague. The blood vessels in arms and legs become necrotic (possibly also an effect of LPS), and this produces purplish blotches on the skin that may have given the disease the nickname "black death". (This name was introduced centuries after the major European epidemic, and as an alternative explanation could have resulted from a literal translation of Latin *atra mors* which can mean "terrible death" as well as "black death".)

Outbreaks of plague were not unheard of in Ancient Europe. The sixth plague described in *Exodus* (following the one possibly caused by anthrax) could have been an outbreak of *Y. pestis*, as well as other plagues mentioned in the Bible. A severe plague epidemic started in Egypt in 542 AD during the reign of emperor Justinian and spread over the Roman Empire, killing thousands in Europe, Northern Africa, and Central Asia. Before that, the Athenian plague that raged during the Peloponnesian War (430 BC), when the city of Athens was overfilled with people, might have influenced the outcome of their war against the Spartans, since it killed their leader Pericles in a second wave of the disease. As with all these epidemics, scholars disagree which pathogen caused it, but *Y. pestis* is among the possible candidates.

The most infamous pandemic (which is an epidemic spanning multiple countries or continents) occurred in the fourteenth century. It may have started in Mongolia (China) and traveled over land to Mesopotamia (Iraq) and Asia Minor (present-day Turkey). It entered Europe via traders whose ships landed in Messina in Sicily (Italy) in 1347. From there, it spread rapidly, as this must have been a particular virulent strain of *Y. pestis*. The disease followed trade routes via Sicily, Genoa, and Venice, all through Europe, Northern Africa, and the Balearics, traveling with the infected rats, fleas, and men en route. It most frequently hit coastal areas, as ships were the most common modes of long-distance transportation. Sequential waves rolled over Europe, waning in winter when the rat's fleas were dormant, to be back next spring. By the end of the pandemic, it is estimated that between one-quarter and one-third of the European population had died. It caused long-lasting social changes, as cities collapsed, workers became rare so that wages increased, faith in the Church was damaged, and the recovering society embraced cultural developments that started the Renaissance.

Since then, *Y. pestis* has remained an unwanted visitor. Epidemics have been recorded in the following centuries, and although none was as severe as that of the Black Death episode, an epidemic in England in 1665 took tens of thousands of lives. A large epidemic in China that started in 1860 resulted in the first isolation of the responsible pathogen, in 1864 in Hong Kong, by Alexander Yersin (1863–1943), who is remembered in the name of the genus *Yersinia*. In the early twentieth century, an epidemic in India killed approximately 10 million people, and an outbreak during the Vietnam War caused over 4000 cases. Nowadays, cases of bubonic plague are rare in places where rat populations (and their fleas) are kept in check. Between 1000 and 3000 cases are reported yearly, mostly from Russia, the Middle East, China, and other parts of Asia. Madagascar, parts of Africa, and Brazil also report regular cases. In the United States, 10–15 cases are recorded yearly, especially in hunters or individuals exposed to wildlife.

There are striking differences between these recorded and well-documented outbreaks and the medieval Black Death catastrophe, both in lethality (there were far more lethal cases in the medieval epidemic than in later ones) and in the speed with which the disease spread (it could spread between cities in days, which is faster than a rat or flea can travel). Moreover, the disease progressed much more rapidly, if we believe historical records, than bubonic plague typically does. People were recorded to be healthy at breakfast and dead before sunset. This has shed serious doubt on whether *Y. pestis* was really the case of the Bleak Death. In an attempt to solve the case once and for all, bacterial DNA was recovered from the teeth of fourteenth century burials in Montpellier, France, that were believed to have been victims of the epidemic that has taken such a large toll of the French population. By PCR it was shown that the DNA belonged to *Y. pestis*, but this will not be the last word, and the debate continues. The latest insight came when it was demonstrated that human lice might have carried the infection from one person to the next, traveling with the speed of (fleeing) man, rather than rats.

In 1994, panic broke out in the districts of Gujarat and Maharashtra in India, when cases of plague were identified in the cities of Surat and Beed. Rumor spread that a plague epidemic was developing. Nurses and doctors left their hospitals as scared as other inhabitants who fled in panic from the much-dreaded disease. The authorities tried to calm the public but failed because of the lack of trust, the lack of information, and an urge to protect their reputations, leading to denial policies. Two versions of the episode exist. In one view, the epidemic lasted only two weeks and was stamped out by effective measures of treatment and diagnosis, whereby 1200 persons got infected and a number of these died. *Y. pestis* had been shown to be present (by PCR analysis) in a number of individuals as well as in locally caught rodents. A few bacterial isolates were available for confirmative testing. In another view, there had not been any plague. The diseased had suffered from a collection of other infections, and the PCR method had been nonspecific. The bacteriology had been faulty, and the isolation of the pathogen was not evidence for an outbreak but instead pointed to the normal presence of this bacterium that is endemic in the region; panic and misinformation had resulted in an "epidemic" that never was. Both views can be politically tainted. To deny the presence of plague can be desired by those who fear the negative consequences and reputation damage due to the disease. To praise the effectiveness of local authorities in keeping such a dangerous disease under control can be favorably viewed by others. Even for an outbreak occurring under our nose, it can be difficult to obtain reliable information.

A close cousin of *Y. pestis* is *Yersinia pseudotuberculosis*. This pathogen causes a far milder disease: gastroenteritis, which is the medical term for food poisoning.

From comparison of the genes and genomes of these two cousins, it was concluded that *Y. pestis* evolved from *Y. pseudotuberculosis*, by the uptake of a plasmid (a DNA molecule much smaller than a chromosome, but replicating independently) on which its many virulence genes are located. Unknown is which organism donated the plasmid to *pestis*-in-the-making, but it could be estimated when this must have happened: between 1500 and 20,000 years ago.

Although this seems like a long time, if correct, it would mean that the plague evolved only after man already existed. It is unlikely, however, that *Y. pestis* evolved in humans. It most probably originated in rodents, which are still its most important host. *B. anthracis* and *Y. pestis*, these much-dreaded pathogens, are not interested in humans—all they need is a host to survive and multiply in, and that unfortunate host is in most cases an animal.

Even the bacteria causing cholera can be associated with animals, although this is less generally known, and the animals live in sea rather than on land. Cholera is caused by ingestion of the *V. cholerae* bacteria, mostly through water (or sometimes food) that is contaminated with feces, resulting in intestinal colonization. The bacteria then start to produce their potent toxin, which was already discussed in Chapter 9. It is this toxin that is responsible for the diarrhea. We know this because *V. cholerae* strains in which the toxin genes had been inactivated (a common approach to determine the function of genes, as explained in Chapter 11) were no longer able to cause cholera. In fact, many *V. cholerae* strains are naturally unable to cause cholera, despite their name, since they do not have the toxin genes. These harmless bacteria live mostly in the sea, where they are associated with copepods (small crustaceans) or plankton and are even found inside fish. *V. cholerae* may spread through sea birds feeding on these crustaceans and fish. It is likely that toxin-carrying *V. cholerae* strains share the same ecology. Indeed, occasionally people get infected from consumption of contaminated fish. When such sporadic cases occur in areas where sewage and drinking water are not strictly separated, an epidemic can rapidly develop.

The association of cholera with (contaminated) water was first discovered by John Snow (1813–1858), a British physician. During a cholera outbreak in London, he famously performed an epidemiological investigation avant-la-lettre, concluding that many victims had one factor in common: a public water pump on Broad Street. In contrast to most of his contemporaries, he was already convinced that water-borne germs rather than "foul air" were the cause of infection, which put him on the right track. He specifically investigated which water had been used by households with a cholera victim. He also plotted lethal cases on a map of London to show their clustering around the culprit pump, much like we still do in present times to identify an epidemic relationship between sporadic cases of an infection with a common source. That approach was not uncommon in his time, but he had difficulty convincing others of his ideas. He investigated the water from the pump with his microscope but could not conclusively identify any germs. Nevertheless, he made the local authorities close the pump, which in hindsight was a proper measure.

As the London outbreak demonstrates, cholera once occurred in Europe, as it did in Northern America. Improved sewage to separate wastewater from drinking water has extinguished the disease from these continents, but cholera is still endemic in southern Asia and parts of Africa and Latin America, where it appears in coastal areas with seasonal regularity. Cholera victims who survive the disease are immune to the strain they had suffered from, so that the next time cholera appears (in countries where it is endemic it often coincides with yearly flooding), these individuals are protected. Most victims in these endemic regions are therefore children who have not yet been exposed. However, if *V. cholerae* spreads to novel regions (at the moment migratory sea birds that have eaten infected copepods or fish are suspected to spread the disease over long distances, although human traffic is sometimes also to blame), it hits a population that is immunologically naive, as nobody is immune yet. A fierce epidemic can result, making many victims among children and adults alike.

At least seven pandemic waves of cholera have occurred since the first recorded pandemic of 1817. The strain causing the current seventh and longest pandemic is called *V. cholerae* El Tor and first appeared in Sulawesi, Indonesia, in 1961, although most other pandemics started in India in the Ganges delta. It was during the second pandemic, which reached Britain in the late 1830s, that John Snow made his observations. The same pandemic reached Canada by ship, traveling with infected Irish immigrants. The third and fourth pandemic soared in the United States in the 1850s and 1870s, whereas the fifth pandemic hit Latin America hard. It was during this fifth pandemic that the "comma bacilli" were first discovered, by Robert Koch, from Egyptian patients in 1883. The sixth pandemic lasted from 1899 till 1923 and was mostly confined to the near and middle East as well as the Balkan Peninsula. Cholera had been absent from Latin America for a century, as the sixth pandemic never reached the continent, when it suddenly reappeared in 1991, during the seventh pandemic. This was not only the longest but also the most widely spread pandemic, although it was less lethal than previous pandemics. Nevertheless, modern cholera takes its toll. A very large epidemic among Rwandan refugees in Goma, Zambia, took 12,000 lives in 1994, and Zimbabwe suffered from an epidemic in 2008 that involved 100,000 patients and may have caused 5000 deaths. The recent earthquake in Haiti (January 2010) was followed by a cholera epidemic 10 months later, whereas the country had been free from cholera for a century. The complete genome sequence of *V. Cholerae* has been established, and it revealed that this bacterium contains two chromosomes, meaning that its genes are divided over two large molecules of DNA. Although many bacteria have only one single chromosome, there are bacteria that beg to differ, and *V. Cholerae* is one of them.

This short and incomplete list of recent epidemics shows that the disease mostly hits countries that are already suffering from natural or man-made disasters. Sadly, most of these cases could have been prevented if clean drinking water had been available and proper sanitation had ensured that food would not get contaminated. Of the three diseases, anthrax, pest, and cholera, the last nowadays takes the most victims, although it is well known how to prevent it. An oral vaccine that would be able to keep epidemics under control has been developed, and this could be protective also in developing countries, but it is not widely used. One problem is that the vaccine works only after two doses, which have to be taken one week apart, and the other is that it must be taken with clean water, difficult requirements in a crisis situation. Moreover, the disease hits unexpectedly in areas where it is not endemic, so policy makers and health care workers are unprepared. Nevertheless, vaccination is probably the way forward to prevent large outbreaks in the future. It can only be hoped that one day we can state that cholera, like pest and anthrax, is mostly a disease of the past.

20

Our Bacteria

To counter the previous chapter on bad bacteria, this one will deal with benign species. It cannot be denied: humans are mostly a scaffold for the bacteria that live in and on us. The cells of our body are outnumbered by a factor of ten by the number of microbes that we harbor. The only reason why this remained unnoticed for so long is because our body cells are so much bigger than the bacterial and archaeal cells that can be considered "part" of our being; our microflora only contributes about 1.5 kg to our weight, despite their numerical excess. Our internal organs are usually free of microbes as long as we do not suffer from an infection (ignoring the odd virus that can hide in our cells without causing symptoms) but the surface of our body is covered with bacteria. The digestive tract counts as a surface too, as it is in contact with the outside world through the food we digest. All our surfaces that can encounter bacteria are either covered with skin or epithelium, the latter sometimes covered with a layer of mucus.

Some of the bacteria living on our surface can be a bit of a nuisance. The armpit, which contains a large number of sweat glands, is home to bacteria that transform odorless sweat into smelly volatile substances. The dominant bacteria living here are *Staphylococcus* and *Corynebacterium* species, whereby the latter (which belong to the phylum of Actinobacteria) are the most important producers of smelly chemicals. Similar aerobic organisms may be responsible for smelly feet, but this has not yet been subject of vigorous research.

Only a fraction of the bacteria we share our lives with have been characterized. From metagenomic analysis (the sequencing of isolated bacterial DNA from

Bacteria: The Benign, the Bad, and the Beautiful, First Edition. Trudy M. Wassenaar.

a sample without prior culturing of bacteria; see Chapter 13), we know that approximately one hundred times more bacterial species live in our gut, in our mouth, or on our skin, than can be cultured. Of the species whose presence was shown by their DNA, approximately 400 live in the mouth, and 1000 in the intestine. Some of the oral flora can also be found in the intestine, but, by and large, bacteria display a preference for the site they occupy. Even within the gut, the flora of the small intestine differs from that of the colon (where the numbers are the highest). The number of bacterial cells present at a given site is, of course, much higher than the number of their species. Estimates report that 100 million times a million (10^{14}) bacterial cells are living in the gut, which makes it one of the most densely populated known microbial ecosystems on earth. Our skin harbors one hundred times fewer cells, and again one hundred times fewer bacteria reside in the mouth. The number of genes carried by all of our bacteria together by far exceeds the number of genes our own cells contain. Every human cell has the same set of approximately 23,000 genes, whereas the complete *microbiome* (the combined genomes of all our bacteria) of one individual easily adds up to a million different genes. Together with our prokaryotes we form a functional unity, as all these cells and their genes may interact with each other.

The ongoing Human Microbiome Project aims to sequence all these bacterial genomes following the metagenomic approach, so that even those bacteria we cannot presently culture can be genetically characterized. Since the intestine is the most heavily populated organ, this is the first site to be explored here. The findings have so far identified Gram-positive Firmicutes and Gram-negative Bacteroidetes as the phyla that are mostly distributed in the gut. These two phyla may cover 80% of the bacteria living in the gut. Typical Bacteroidetes genera are *Bacteroides* and *Prevotella*; the Firmicutes are usually represented by *Clostridium*, *Enterococcus*, *Lactobacillus*, and *Ruminococcus* species. The Gram-negative Proteobacteria, to which *E. coli* belongs, are only minor players in a healthy adult gut, together with the Gram-positive Actinobacteria (e.g., *Bifidobacterium*). Note that *E. coli* was described in Chapter 9 as a pathogen, producing various toxins to harm its host. Nevertheless, *E. coli* strains are usually found in the gut of healthy individuals, and these strains are completely harmless. The difference is due to their genes; different types of *E. coli* display a wide variety in their gene content (and even in the number of genes they possess). Depending on the type of *E. coli*, these bacteria are either beneficial or pathogenic.

Compared to other ecosystems where diverse microfloras thrive, such as terrestrial soil or coastal sediments, the human gut contains fewer bacterial phyla, but within those phyla represented, the diversity of genera and species is impressive. At a broad level, the distribution of the different phyla and the main groups therein is more or less constant, both between individuals and within an individual over time. However, when we zoom in on genera and species, their distribution varies extensively between individuals and is not constant over time. That variation is not greater in siblings than in identical (monozygotic) twins, suggesting that there is not much of a genetic influence in the development of the microflora. At the same time, the differences observed between family members is less extensive than that between unrelated individuals, which hints that cultural, social, and dietary factors, or a combination of these may play a role. It seems that every person collects his or her own microflora over time, but there is not a "core" of species always and predominantly present in every individual. At best, we can identify a collection of species that are likely to be present in most persons.

The picture changes, however, if we consider the bacterial genes present in the gut microbiome, irrespective of which species carry these genes. Now we can recognize a conserved collection of genes that are always present, which builds the "genetic core" of the intestinal microbiome. Apparently, it does not matter which species contributes to the community, as long as particular genes (and the phenotypes these genes produce) are represented. The community truly acts as a whole, where the individual species is less important than the synergy produced by all members working together. The population of our microflora could be considered a "superorganism" whose members may vary, but which performs its task more or less constantly. Apart from the conserved core genes, there are numerous genes that can, but do not have to be present, depending on the species that are residing, and whose presence might vary over time in an individual.

A baby develops in a sterile womb, and will collect the bacteria that colonize various sites quickly during and after birth. A natural birth is a good way to receive a large variety of vaginal microorganisms from the mother, while a diet of breast milk favors particular species to multiply predominantly in the gut of a newborn. *Bifidobacterium* species are dominant colonizers of a young gut, and are considered healthy and beneficial even later in life. Their cells have a typical "Y" shape. They may remain present all through old age, but in much lower numbers than in babies. Babies delivered by Caesarean section see their first light in an environment with a low bacterial load; their initial gut microflora less resembles that of the birth canal than that of the mother's skin, but eventually the differences mostly disappear. The introduction of solid food introduces a major change in the intestinal microflora, and by the age of two, the microflora is more or less developed and not significantly different from that of an adult.

The gut microflora seems to undergo extensive changes again as we age. The gut of the elderly contains more *Ruminococcus* and fewer *Bifidobacteria* and *Eubacteria* (the latter are Firmicutes). However, different findings have been reported depending on the geographical regions where the investigated individuals lived. For these studies, middle-aged and senior healthy persons are usually compared. The techniques applied are still novel and individuals have not yet been extensively followed over time, so that the wide diversity reported for the microflora associated with the elderly may be partly due to individual variation, rather than an increase in diversity over time within a person. In general, though, it seems the gut microflora becomes less stable and more diverse with increasing age. The reason for this could be changes in diet, a weaker immune system, and increased medication as age increases.

When the gut microflora of obese and lean people were compared, substantial differences were detected: obese individuals had more Firmicutes and fewer Bacteroidetes in their gut than lean persons. But after the overweight participants had lost weight as a result of a low-fat diet, their ratio was restored to that of the thinner group. It could be that obesity provides conditions that result in a particular Firmicute-rich microflora. Alternatively, it may be that different bacterial distributions cause obesity. Or possibly, obesity and a changed microflora are both the result of an external factor, such as a high-calorie food intake. There is still discussion whether obesity is cause or effect of the variation in species and genus profiles and shifts between phyla, but some data indicate that a difference in gut microflora may already exist during infancy. Whether this causes obesity or is an (indirect) effect of other factors that eventually lead to the condition remains a subject of debate.

For those who are scared of bacteria, and wash hands excessively, the desire might exist to get rid of these bugs completely, but that would obviously not be a good idea. Animals, including humans, have evolved with their microflora. Nevertheless, an animal can survive without bacteria; laboratory mice can be made virtually germ-free by extensive treatment with strong antibiotics, and they can even be bred completely sterile inside an isolator. Mice that are born germ-free do not die, but their intestines swell up and do not develop properly. On a special (sterile) diet, these animals survive, and are used as a model to investigate the role of the intestinal flora on their development and health. It turns out that the microflora has a large effect on both and the immune system plays an important role in the interaction between the host and its microbes.

The immune system, already introduced in Chapter 5, is constantly on the alert to identify and destroy pathogens that enter our body, and is particularly active in the gut where pathogens might enter via food. At the same time, it must leave the normal inhabitants residing there in peace. The gut contains a large number of immune cells—together with the skin, it is the largest immune organ of the body—and all these cells must know which bacteria to leave alone and when to spring into action. Sometimes, this regulation goes awry, and an inflammatory reaction is raised against bacteria that are completely harmless. This causes harm that the body does to itself, and is probably at the basis of inflammatory bowel disease (IBD) and Crohn's disease, two common chronic diseases of the gut. The developing microflora of an infant teaches the immune system what should be tolerated and what not. If this goes wrong, an autoimmune disease can result. A lot of research is currently dedicated to determine the function of the intestinal microflora and development of asthma, eczema, and a number of other atopic (allergy-associated) diseases. Even though the intestinal microflora only communicates with immune cells in the gut, once these respond, they can elicit reactions in other parts of the body, too, as was already seen in Chapter 5.

The idea that health can be improved by adding "good" bacteria to the gut is already one hundred years old. *Probiotic* bacteria, as they are collectively known, are supposed to improve health by improving food uptake, by soothing the immune system (to avoid allergies), by activating it (to combat infections), or by competing with pathogenic bacteria so that the latter have no chance to proliferate. A lot of research, carried out to investigate how well bacteria perform all these tasks, concentrates on species that have traditionally been used for fermentation of food and dairy products. Since these are known to be completely safe for consumption, they were chosen as top candidates to function as a probiotic: species belonging to *Lactococcus*, *Lactobacillus*, *Enterococcus* (all Firmicutes), *Bifidobacterium* (Actinobacteria), and a couple of other genera. However, these bacteria are good at fermentation, but may or may not excel as health promoters. They were mostly chosen because they were safe to use, but whether they are most suitable as a probiotic is not exactly clear. Considering the large variety of species typically present in a gut, there must be hundreds of species that can potentially promote health. Maybe they only do so when they are plentiful and varied, in other words, when there is sufficient diversity. The idea that one species (and one strain thereof) can make a difference in such a complex system as the gut microbiome is maybe a bit naive. Moreover, the target individuals who would benefit from probiotics are a highly diverse group of newborns, children, adults, or the elderly, that may be healthy (but aiming for even better health) or sickened (suffering from an array of different diseases). One magic bullet for the diverse applications for which probiotics are being developed can hardly be expected. Nevertheless, there seems to be some truth in the general idea that particular species can be beneficial, and the way they interact with the immune system is quite well understood.

So far, we have dealt with the intestine, where every person carries his or her own individual combination of bacteria (including archaea). This individual load also applies to the bacteria living on the skin, where in some locations 10 million bacteria may live on a square centimeter. When the bacteria present on the hands of 51 students were investigated, by sequencing of the ribosomal RNA genes present, each of the students was found to carry a unique set of species. It was demonstrated that bacteria living on a right hand are different from those of a left hand, even if these two hands belong to one person. On average, 150 different species live on a hand palm, belonging to more than 25 phyla, but three phyla predominated: the Actinobacteria, Firmicutes, and Proteobacteria accounted for 94% of all bacterial DNA detected. Five genera were found to be predominant on most of the students' hands: *Propionibacterium* accounted for nearly one-third of the findings, *Streptococcus* for nearly one-fifth, and *Staphylococcus*, *Corynebacterium*, and *Lactobacillus* made up smaller fractions. All of these genera are known to be common inhabitants of skin from growth-dependent observations (i.e., the bacteria had been cultured). But that the diversity at the species level, detected here in a growth-independent manner, was so extensive was a novel discovery. Any two hands only shared 13% of species, and the left and right hand of an individual were identical for only 17% of all bacteria detected. The species that dominated on male students' hands differed from those on the hands of females, and the latter displayed greater diversity for their hand microflora. The investigated students had not been asked to refrain from hand washing: they were asked to be tested after they left an examination room. Sweat production (regulated by anxiety) influences the skin microflora, but it was not disclosed which examination the students were taking nor were the examination results correlated to the bacterial findings.

Could one expect a correlation between the examination results and the bacteria living on the students' hands? Probably not, but recent findings have suggested that bacteria can have an effect on brain activity, including emotions and memory. In order to do so, the bacteria need to communicate with the immune system, and so far it has not been demonstrated that bacteria living on the skin are detected by the immune system. The first hints that bacteria may have mood-altering activities came from observations that the immune system can modulate emotions; for instance, chronic immune diseases can lead to depression. Activation of the immune system by bacterial products can influence both memory and the ability to learn, at least in laboratory mice. By injecting mice with (killed) *Mycobacterium vaccae* bacteria, the animals could better endure mild stress, as they produced higher levels of serotonin in their brain. Such findings were unexpected, since for a long time it had been thought that the brain was completely separated from immune cells. Since then, there have been cases described where bacterial infections were at the basis of sudden personality changes. Take, for example, the case of a 12-year-old boy who suddenly developed obsessive compulsive disorders and became dysfunctional within six weeks, although he had lived a completely normal life up till then. The symptoms started to improve when he was prescribed antibiotics to combat a *Streptococcus* infection. After six months of treatment (far longer than a normal course to fight this type of infection), the boy was back to normal. Some sufferers from Tourette's syndrome (a compulsive disorder with tics and sudden vocal outbursts) could also improve from antibacterial treatment against *Streptococcus*—although certainly not all Tourette's cases are due to this bacterium, a normal inhabitant of the throat. The communication that takes place between gut bacteria and our immune system was already pointed out, and it can thus be hypothesized that certain gut bacteria contribute to happiness, while others may cause sadness. Unfortunately, we do not yet know how which species do what, so we cannot improve our bad mood with a simple probiotic drink. Who knows, though? Maybe one day happiness will be contagious, and kissing a happy person may bring on good feelings for more than one reason.

A lot of work has been carried out to identify the oral flora, and its implications in caries and periodontitis. Dental plaque, the bacterial growth on teeth and in the gingival cavities that can lead to caries or gingivitis, is now recognized to be formed by complex biofilms of multiple bacterial species. An oral biofilm typically starts with attachment of Gram-positive bacteria (for instance, *Streptococcus* species that belong to the Firmicutes and *Fusobacterium* species, which have named their phylum Fusobacteria), to which Gram-negatives later join. Dental plaque does not always lead to disease, which might depend on the species of bacteria that are present. Notably, the presence of *Porphyromonas gingivalis*, *Tannerella forsythia* (both belonging to the phylum Bacteroidetes) or *Treponema denticola* (a Gram-negative Spirochete) is associated with disease. However, many persons carry these species in their mouths without any adverse effect, so obviously dental disease is the outcome of multiple factors, of which the presence of "pathogenic" bacteria is only one.

Oral malodor (halitosis is the medical term) is caused by anaerobic bacteria, which can live on the tongue or be the cause of a chronic (as in periodontitis) or acute (gingivitis) infection. Other conditions that are not due to bacteria can also result in bad breath. When bacteria are involved, apart from those mentioned above, *Prevotella intermedia*, *Fusobacterium nucleatum*, and *Eubacterium* species can be implied (the latter are a species of Firmicutes, not to be confused with the general term eubacteria, for all prokaryotes that are not archaea). The condition is very common, but a self-perceived "bad breath" is not always noticeable by observers. The opposite occurs as well, in that halitosis is not always known to the carrier.

The mouth is in open connection to the larger nasopharynx as well as the throat, each of which harbors specific bacterial inhabitants. To avoid a rather boring summation of findings, we will leave these alone and continue with some more unexpected bacterial visitors.

Although the internal organs ought to be sterile, this is not always the case. When artificial hip prosthetics had to be replaced (because they had come loose, or got infected), microbiologists investigated the metal joints after they were taken out of the patient. They investigated the presence of bacteria both by culture and by sequencing their telltale ribosomal RNA. Much to their surprise, even loosened, but noninfected (or so they thought) joints were covered with bacteria. By culture, seven different species were found to live on the joints, of which *Leifsonia aquatica* (previously called *Corynebacterium aquaticum)* was most often found. This is a Gram-positive belonging to the phylum Actinobacteria with a high G-C genome that is not frequently associated with humans—although it is known to cause opportunistic infections. It was first isolated from distilled water, and is generally found in water; it can cause disease in sugarcane. How it got on the surface of a prosthetic hip inside a number of healthy individuals could not be explained. By DNA analysis, the most frequently found species on the prosthetics was *Lysobacter enzymogenes*; these are Gram-negative Proteobacteria that are also more frequently associated with plants than with human patients, although again, opportunistic infections caused by these bacteria have been described. This species produces enzymes and antibiotics of biotechnological interest. The most unusual finding was the DNA of a thermophilic bacterium that normally lives in hydrothermal vents. How it got inside the hip of a human being, and whether it could multiply in this unusual environment remains unknown.

Most of the bacteria described in this chapter do not cause disease, although we share an intimate relationship with them. If people were more aware of the normal microflora that is part of us, there might be fewer cases of excessive hygiene, which can develop into obsessive compulsive disorders. Even when practiced nonobsessively, hygiene may not always be a positive thing. One theory states that a lack of exposure to bacteria in general, or pathogens (including parasites) in particular, is the cause for the increase in allergy-related diseases that seem to be a consequence of a Westernized lifestyle. This *hygiene hypothesis* was first proposed in 1989. Allergies, diabetes, and other immunological diseases are far more common in developed than in developing countries, and this correlates with a lower incidence of infectious diseases in the first. However, to prove that these medical conditions are the result of lower exposure to bacteria or pathogens turns out to be difficult.

Immigrants from developing countries who came to live in industrialized countries were protected against allergic, "Western" diseases, but their children who grew up in their new nation were not. Still, such epidemiological evidence is not pointing to a cause—it can only identify correlations. Experimental studies with animal models have provided evidence that frequent exposure to infections indeed induces long-term changes in the immune response of the body, which would fit in with the general hygiene hypothesis, but translating animal results to the human host is tricky. Moreover, a number of observations remain hard to explain. The increase in immune diseases is continuing in present times, whereas the low level of infant disease in Westernized countries is no longer significantly decreasing. The most significant decrease in infant disease occurred in the middle of the last century, whereas the increase in (children's) allergies seemed to happen decades later. Particular infections can even be the cause of autoimmune disease, instead of preventing these. The atopic (allergy-related) diseases that would be caused by lack of bacterial exposure are a moving target, with type 2 diabetes as the latest addition, to an existing array of eczema, food allergies, hay fever, asthma, or IBD. The presumed decreased exposure that would explain the mechanism behind these diseases is also constantly being redefined. At first, a lack of exposure to pathogens in general was proposed, but now a lack of intestinal microbial diversity is being proposed, or else a lack of worm infections is considered as a cause. The hygiene theory has not matured yet, and at the moment the jury is still out to decide how valid the theory proves to be.

The human body, its microflora, and the interactions between these two remain fascinating subjects of research. Scientific findings in this field frequently reach the press headlines. We can expect to hear more on this topic in the next few years, but it seems that one major discovery is bound to be made: our microflora and the interactions between bacteria and our body cells will turn out to be even more complex than we thought.

21

Sensing Bacteria

Which of our senses can detect bacteria? Obviously, we cannot feel the crawling millions that inhabit our skin, mouth, and other "external" body parts. The only time we indirectly feel the presence of bacteria is when they cause an infection, which is usually accompanied by pain. We cannot see them either, without the help of a microscope, at least when they are few. But large populations of bacteria are clearly visible. The water of a flower vase will turn cloudy over time because of bacterial growth. When growing on a solid surface, for instance, on an agar plate, bacteria will from colonies. A colony is the offspring of a single cell and is visible as heaps of various shapes and colors, depending on the species. Some social species produce more daunting patterns when food is scarce, as we have seen in Chapter 17. Typically, a colony becomes visible after incubating an agar plate at 37°C for 10–12 h, when about a million cells have formed from every single living cell that was streaked on the agar, and these million-cell heaps are visible as colonies. But what is typical in the diverse bacterial world that has been described on these pages? The bacteria forming colonies at that temperature in a laboratory are mostly human-associated bacteria, frequently pathogens growing in a medical microbiology laboratory (as that is where a lot of laboratory culture is carried out), and often a member of the Proteobacteria or Firmicutes. Other bacteria may grow much faster, so that their colonies would become visible within hours, such as *Clostridium perfringens*, which won the fast-growing contest in Chapter 16. Others, such as *Mycobacterium tuberculosis* grow so slow that a colony appears after weeks only. Many others would require different temperatures, or even high

Bacteria: The Benign, the Bad, and the Beautiful, First Edition. Trudy M. Wassenaar.

pressure chambers to form colonies, if we can make them grow in a petri dish at all.

We can smell bacteria from the gases that they produce. Some bacteria produce hydrogen disulfide (H_2S), a gas that smells of rotten eggs or sewage (which smells like that because of the bacteria living there). Our nose is very sensitive to hydrogen disulfide. We can smell it at extremely low concentrations of a few parts per billion. Another badly smelling compound produced by bacteria is butyric acid. It is a product of anaerobic fermentation. Humans can smell butyric acid at 10 ppm, but dogs can detect concentrations a thousand times lower than that. It is produced by some *Clostridium* bacteria in the gut, where the gas has an antiinflammatory activity, and together with H_2S, it contributes to the foul smell of flatulence. Sweat starts to smell only after bacteria have degraded it to propionic acid (which smells a bit like vinegar) and isovaleric acid. The smelly armpit culprits are mainly *Propionibacterium* species and *Staphylococcus epidermis*.

Obviously, we can taste bacteria, at least we taste their works after they have grown in cheese, yogurt, sauerkraut, and many other foodstuffs that are prepared with, and receive their specific taste from bacteria. The stale taste in your mouth after a night of sleep is also the result of bacterial activity, which have multiplied without the flushing action of saliva.

But we cannot hear bacteria. They do not produce air vibrations strong enough for any ear to be picked up.

Bacteria have senses, too. They sense their outside in order to react accordingly, which was one of the requirements of life (see Chapter 6). Many bacteria respond to temperature. Especially, dangerously high temperatures are quickly registered. Whether a temperature is dangerously high depends on the species. Psychrophiles may find 10°C unbearably hot, whereas thermophiles may start feeling comfortable at 80°C or more, but every bacterial cell can be damaged by temperatures that are too hot for them to endure. One general mechanism how bacteria sense such dangerous temperatures is by detecting their temperature-damaged proteins. When protein is heated, the molecules change shape, but the temperature at which this happens depends on the kind of protein. For instance, the various kinds of albumin proteins found in egg white solidify at temperatures between 61 and 84°C, a fact that is used in molecular cuisine. Some proteins in a bacterial cell will start to change shape (and as a consequence will malfunction) when they get too hot, and the cell contains low amounts of chaperone proteins that notice this change of shape. Chaperone proteins were mentioned already in Chapter 5, where they were introduced as proteins that embrace other proteins to keep these in a desirable shape. Some chaperones, the so-called *heat shock proteins*, bind to any protein that is deformed because of heat (or as a result of a number of other stresses), with the aim to refold it into its normal, functional shape. When sufficient heat shock chaperones are bound to damaged protein, the concentration of free chaperone is decreased, and as a result, the cell starts to produce more of it. This makes sense, because when the cell gets too hot, more and more protein will get damaged, and only then a lot of repairing heat shock chaperones is necessary. The *upregulation* of heat shock proteins, as the process of suddenly producing extra protein is called, can occur very rapidly after a critical temperature is reached, in the order of minutes only. Protein that is heat-damaged beyond repair will be degraded. Together with an increased production of heat shock chaperones, a number of other processes change rapidly in the cell as a result of non-permissive temperatures, for instance, cell division will be switched down. This collective change of cell activity is called a *stress response*, and it can be the result of heat or osmotic stress, as well as other damaging stresses, for all of which chaperones are key initiators. Bacteria do not have a thermometer, but they do notice it when it gets hot. Apparently, this is a very ancient protective mechanism, because the genes for heat shock proteins are strongly conserved in a wide variety of bacteria.

Bacteria also register what chemicals are present in their environment, for which they have sensors on their outside. These can be very sensitive, as some can sense a change of only 0.1% in the concentration of the compound to which they react. The sensors are dedicated proteins that stick in the membranes, looking out on the outside. These sensors can amplify infinitesimal changes in the environment, which they "report" to the inside of the cell in an amplified form. Not only do they notice very small changes in concentrations but they are also able to do this for concentrations that may differ five orders of magnitude. There are no other known systems in biology that are so sensitive over such a wide range of concentrations. The magnification of the signal is the result of many sensors acting in cooperation. When one sensor molecule has caught a sugar molecule, for instance, that was not there before (an indication of an increase in concentration), it slightly changes its charge and sometimes its shape in the membrane, and by doing so, neighboring sensors do the same. All of these sensors now produce a signal inside the cell, so that the effect of one sensor protein binding one sugar molecule is amplified into multiple signals.

What signal can a sensor protein send to the inside of the cell? Phosphates are the key. Every sensor protein has two functions: apart from binding its preferred compound on the outside, it is also an enzyme, as it can glue a phosphate onto itself on its protein part that faces the inner side of the membrane. Enzymes that can attach a phosphate group onto something else are called *kinases*. The sensors are autokinases, as they phosphorylate themselves, after their outside has sensed the correct signal. Kinases and their antipoles phosphatases (which splice phosphate groups from substrates) are regulators of signals in cells, not only in prokaryotic cells but also in eukaryotic cells. But the combined function of sensor and autokinase activity in one protein is unique to the bacterial world. The signal does not stop here. Once the autokinase has charged itself with phosphate, it will transfer this potent group to another specific protein called the *response regulator*. Every autokinase has its own response regulator to which it will donate its phosphate once it has sensed the correct stimulus outside.

The addition (or removal) of a phosphate to a particular site of a protein (often an enzyme) will change its charge and sometimes its shape and, as a result of that, its function. The phosphate that was added to the response regulator causes this protein to do something it did not do before, for instance, bind to a specific location on the DNA. That binding could result in a gene being switched on, so that the enzyme needed to process the sugar (which was what the sensor had detected in our example) is produced, and the sugar can be degraded once it enters the cell. This way, the bacterial cell prepares itself for a new food type: it has changed its response to a changed environment. The results of protein phosphorylation can be very diverse, depending on the kind of response regulator being activated. An alternative phosphorylated response regulator could maybe activate the flagellar motor so that the bacteria start swimming in a particular direction. In that case, multiple copies of the sugar sensor on the cell membrane will send signals when the sugar concentration slightly increases or decreases, and this will direct the cell toward the food (or away from a compound that is a repellent). Compare it with the children's game where a blindfolded child has to find a treat, and surrounding kids call "hot" or "cold" when the child moves toward or away from the treat. The movement will be staggering, but overall follow a direction, just like the brisk movement of swimming bacteria.

The official term for the signaling pathway described above, which is unique to the bacterial world, is *two-component signal transduction system,* which is quite a mouthful, even for scientists. The "two-component" part in this term relates to the sensoring autokinase (officially called a histidine kinase) in the cell's membrane and the response regulator inside the cell, whereas the "signal transduction" refers to an original stimulus that is converted into a signal in the cell that has some effect (for instance, switching a gene on).

Since two-component signal transduction systems are unique to bacteria, they can be targets for antibiotics, because, as was explained in Chapter 12, antibiotics should not be toxic to the host cells; a drug targeting a system that host cells do not have is more likely to be safe. Most of the antibiotics developed to target this system bind to the sensory part of the histidine kinase, but since these are very specific, it means developing a novel drug for every pathogen. It would make sense to produce drugs that inhibit the kinase activity of the sensor instead, because that part of the protein is conserved in all sensors. However, an antibiotic with that activity should not inhibit other kinases, which are plentiful in all cells, including eukaryotic cells. Research in this area is currently one of the more promising developments to produce novel antibiotic drugs.

Bacteria sense cold temperatures, too, but they do not do this via chaperones, as is the case when it gets hot. Instead, at colder temperatures, the membrane becomes more rigid, as a result of its physical properties, and this sets a two-component signal transduction system in action: the sensors are this time not activated by binding to a compound, but they react because the membrane has lost flexibility. When the amplification of an external signal is based on cooperation between sensor molecules, one can imagine how they all start signaling when they are all forced in one fixed position, due to increased rigidity of a cold membrane. The resulting cold stress response inside the cell is somewhat similar to a heat stress response, although it started with a different stimulus and the signal was initiated by a different mechanism.

The previous chapter described how our body is home to various bacterial populations that are largely left in peace by the immune system, while the occasional pathogen that tries to gain a foothold must be combated. This requires our mucosal surfaces to "sense" the bad ones and leave the benign ones alone. Recognition of known bad bacteria would not suffice as a defense strategy, as it would mean that any novel pathogen would remain undisturbed. Therefore, the immune system relies on a dual strategy, already introduced in Chapter 5: the adaptive immune response learns from a first exposure and produces specific antibodies that will inactivate the pathogen when it enters a second time; this is how people have lifelong immunity against cholera, when they have survived the disease once, and this is also the way vaccines protect (the vaccine acts as a first-encounter surrogate pathogen). The adaptive immune system is very effective, but the first time it has to react it is rather slow. In contrast, the innate immune system is always ready to react against pathogens that the body has never seen before. This innate system is particularly important at mucosal surfaces.

How does the innate immune system recognize pathogens, and how does it react? Pathogenic bacteria frequently penetrate the mucosal surface and enter deeper tissue (they are *invasive*), where they encounter immune cells. Specialized cells, called *macrophages* (large eaters), will engulf bacteria that they decide are bad. They recognize these by structures on the outside of bacteria that are relatively well conserved, such as flagellin (the protein that builds the flagellar tail), LPS (a component of all Gram-negative outer membranes), or peptidoglycan (a component of all Gram-positive cell walls). Instead of the close match by which antibodies bind to their targets, the receptors of innate immune cells recognize patterns, hence their name *pattern recognition receptors*, or PRRs. The individual flagellin proteins or the exact LPS constituents may differ between different bacterial species, but they all display general patterns that are conserved, and this is recognized by the PRRs on macrophages. A well-studied pattern recognition receptor is the so-called Toll-like receptor, but immune cells rely on a number of recognition receptors to make sure they are well prepared for the occasional pathogen.

Once the macrophage has recognized bad, invasive bacteria with their PRRs, they spring into action. They start secreting messenger molecules called *interleukins* that attract more macrophages, activate the adaptive immune system, and start an inflammatory reaction. The macrophages then "eat" the bound pathogen, which they keep inside their cytosol in a membrane-enclosed vesicle. They then release into this vesicle acids and antimicrobial peptides (very short proteins) called *defensins*. Defensins work like pore-forming toxins: they produce holes in the bacteria's membrane. The production of defensins to combat pathogenic bacteria is a very ancient part of the immune response of animals; even primitive animals produce defensins (although not by macrophages), sometimes as the only means to combat infections. Few bacteria have a chance to survive this macrophage killing, unless they have evolved mechanisms to resist the defensins. Some bacteria alter their cell surface in response to defensins, so that these can no longer harm them.

Salmonella bacteria are especially good in survival within macrophages. They have an important two-component transduction signaling system that responds to an acidic environment as well as to the presence of defensins. These conditions both apply when *Salmonella* is inside a vesicle within a macrophage. As a result of the signal transduction that takes place, *Salmonella* protects itself against the defensins and prepares for a life in the vesicle, where it can not only survive but also multiply.

Although it is ingenious that innate immune cells recognize conserved patterns in bacteria instead of specific structures, it does not solve one problem: how do the cells distinguish between good and bad bacteria, since both types contain flagella, LPS, or peptidoglycan? In fact, they do not. Immune cells that encounter benign bacteria recognize these just like they recognize pathogens. But somehow, the result of benign bacterial recognition is not inflammation, and we still do not understand how that distinction is made. One difference is that benign bacteria do not invade: they leave the cells that form the lining of the intestine intact and do not enter them or slip in between cells. Pathogens can do both, and integrity of the intestinal lining seems to be an important trigger to keep inflammation down, whereas disintegration of the epithelial lining fires all alarm bells.

The interaction between the host and the microbes in the gut, whether benign or bad, is complex and involves two-way communication strategies. Both the host cells and the bacteria respond to each other's presence and signals. They sense each other very well, but our conscience does not register the constant communication going on inside our gut between our cells and our microbes.

22

Bacteria and Mankind

The first humans developed as hunters and gatherers. They must have lived on a protein-rich diet, to enable development of a large, power-hungry brain. Paleontological evidence indeed suggests that fish or meat must have been regular food sources of the early humans. The exploitation of fire that allowed food to be cooked, which increased its nutritional value, may have further supported the mental development of our species. When people prepared and ate meat, either from hunted animals or from prey taken from carnivores, they would have been exposed to the bacteria of the dead animals. Such encounters with animals (dead or alive) impose a risk of contracting zoonotic diseases. A *zoonosis* is an infection that is transmitted from animals to people. Even today, bush meat is a serious risk factor for infectious diseases. The most scary, present-day examples are all viral infections. The hemorrhagic fevers, caused by filamentous viruses such as Ebola, Marburg virus, or Lassa virus result in regular outbreaks in Central or West Africa. Ebola, for instance, can kill apes and humans alike; it has a mortality of 80%, meaning that on average 8 out of 10 infected individuals will die. But the virus infects monkeys without serious symptoms, where it reproduces under the name Simian Herpes B virus. This illustrates that a virus that causes little trouble in one host can wreak havoc in another. (There are also bacteria that live without producing symptoms in one host species, to cause disease in another.) The unaffected host will be a reservoir for the microbe that causes disease in the other, less fortunate host. The origin of HIV was also an animal virus: the simian immunodeficiency virus (SIV), which infects monkeys and apes, is the most likely ancestor of HIV.

Bacteria: The Benign, the Bad, and the Beautiful, First Edition. Trudy M. Wassenaar.

The monkey virus must have infected a person at some stage, maybe during slaughter of a hunted animal, and subsequently adapted to the human host, giving rise to HIV. The first *host jump,* as a change of host specificity for a pathogen is called, of HIV occurred possibly around 1880 in present-day Congo. Several independent host jump occasions, all occurring in Africa, are the basis of the different HIV lineages that can be recognized globally today.

SARS, another virus that threatened to cause a worldwide pandemic, was also of animal origin. The name, short for severe acute respiratory syndrome, was first given to the disease, and subsequently to the virus that was discovered to cause it. It had appeared after a host jump that occurred in China, possibly from raccoon dogs or palm civet cats (we cannot be certain of the natural host of the SARS virus) that are captured and traded alive to be sold as bush meat. The SARS epidemic, which lasted nine months in 2002 and 2003, was fortunately stamped out by international efforts to control and eradicate the disease.

Zoonoses make up a significant proportion of human infectious diseases. Of the estimated 14,000 infectious diseases known today, 600 are shared between humans and various animal species. This is being recognized by policy-makers who promote a "One Health" policy, combating infectious diseases in animals and humans alike. It advocates a better cooperation and collaboration between physicians, veterinarians, and other scientific health professionals, in recognition of the fact that animals and humans share a large proportion of their bacteria. But before we deal with the interaction between humans, animals, and their microbes, we will follow the footsteps of mankind as it explored the world.

According to the "out of Africa" hypothesis, *Homo sapiens* evolved in Africa, from where it populated the world in sequential waves. Various parts of the world were colonized by different, relatively small seeder populations, and some continents, such as the Americas and Australia, were colonized by multiple successive waves. These wandering people would have carried their microflora with them, and passed part of these bacteria on to their children. It is therefore possible to recognize lineages of people not only by their human genes but also by their bacteria. This has been demonstrated using DNA from *Helicobacter pylori*. This Proteobacterium lives exclusively in the stomach of humans, so that the only way for this organism to survive is to colonize humans. Approximately half of all humans are infected. Children usually obtain their infections from their mother, and siblings carry strains that are very similar but not identical, as this organism mutates frequently. In most infected persons the bacteria do not cause trouble, although certain bacterial types exist that are associated with peptic ulcer disease. It is possible to recognize first-generation immigrants from their *H. pylori* because their bacteria more closely resemble those typically found in their country of birth than those of their new country of residence. It has also been shown that large migration patterns from the past can still be recognized.

Three types of *H. pylori* bacteria exist in Africa, consistent with the different people who are thought to have colonized this continent. People who passed from Siberia over the Bering Strait to the New World during the last Ice Age subsequently

populated the Americas, as is still visible in the similarity of *H. pylori* isolated from native Columbians and Alaskans and from people living in East Asia. The later appearance in the Americas of Europeans, and their mixing with Africans who were brought in as slaves, can be recognized in the strains currently carried by American people of mixed European and African origin. Australian aboriginals share their *H. pylori* with the Melanesians and Polynesians, whereas the Australian white population carries bacteria more like those of Europeans. Finally, the Iberian population resembles that of Northern Africa, at least in terms of their stomach bacteria, which is a reflection of the large population shifts that occurred during the Moorish reign. Fewer children are colonized by *H. pylori* nowadays than in previous generations, and the number of infected people is declining. It is expected that the infection will soon become uncommon, which means that such studies on human lineages based on *Helicobacter* carriage, may no longer be possible in the future.

The early hunters and gatherers might have led a nomadic life. The development of agriculture and animal husbandry occurred relatively late in the existence of our species. A fixed domicile might increase the carriage of insect parasites that can hide in shelters, which could favor particular infectious diseases brought on by these parasites (fleas, lice, bed bugs, and the like). But a more important effect on pathogenic load was the close contact with animals that herding and farming has introduced. This has left marks on our genes.

The most obvious effect came from the consumption of milk, which left both a genetic imprint in our genome and introduced new pathogens. Mammals usually consume milk only during the first period of life, and as we saw in the previous chapter, the human intestinal gut flora changes after the introduction of solid food.

The intestinal cells change their metabolic repertoire as well. Adults mammals normally do not tolerate milk (a condition known as *lactose intolerance*), because it contains the milk sugar lactose. In contrast to the young intestine, adult intestinal cells no longer produce lactase, the enzyme needed to degrade this sugar. A mutation in the gene regulating lactase production can change this: as a result of the mutation, the enzyme is now produced in adult intestinal cells as well. Adults who have this mutation digest milk like children do, and milk is an important protein source. In societies where milk-producing animals were kept, the normal condition (lactose intolerance later in life) was outcompeted by the favorable mutation (lactose tolerance). Nowadays, lactose intolerance, once the default of all humans, is seen in 90% of the Asian population but varies from 25% in Sicily to less than 3% in the Netherlands or Scandinavia.

The consumption of milk also introduced novel pathogens in the human population. Brucellosis, a disease caused by various *Brucella* species (Gram-negative Proteobacteria) is associated with the herding of animals, and is spread by consumption of contaminated milk or meat. Cattle can be infected by *Brucella abortus* (as its name suggests, it can cause abortion in infected pregnant cows), pigs by *Brucella suis*, and goats and sheep by *Brucella melitensis*. When humans get infected, flulike symptoms, typically accompanied by excessive sweating and muscle pain, are the result. Without antibiotic treatment, waves of fever that come and go are typical, and during a chronic infection, eventually the bones and joints will be pockmarked with typical lesions that remain visible in the skeleton. After the Mount Vesuvius eruption in Italy (79 AD), the cities of Pompeii and Herculaneum were buried in volcanic ash. The bodies of victims that were excavated nearly 2000 years later still showed the signs of past brucellosis infections. Since cheese made of sheep or goats' milk was part of a typical Roman diet in those days, the disease was most likely caused by *Brucella melitensis*. A fossilized cheese was recovered from Herculaneum in which typical microscopic round (coccal) shapes could still be recognized that might have been *Brucella* bacteria. DNA, however, could no longer be detected, as the hot pyroclastic cloud that had killed and fossilized all living organisms had also destroyed any DNA present in the cheese.

Diseases that leave traces on bones are favorite subjects of research for paleobiologists. The more common enteric and pneumonic diseases from which early humans must have frequently suffered leave no traces on the few remains that are occasionally surfaced, so that their impact cannot be studied. *Mycobacterium leprae* is one of the few bacterial diseases that cause visible deformations of the skeleton. The disease leprosy is still common in many parts of the world, with a quarter of a million patients worldwide. The bacteria survive inside immune cells, just like their cousin, *Mycobacterium tuberculosis*. *M. leprae* is a difficult bacterium to study, as it cannot be cultured in the laboratory; being an endosymbiont, it can only survive inside cells, where it grows extremely slowly. The discovery that these bacteria can infect the nine-banded Armadillo was a breakthrough, as this animal model allowed production of sufficient amounts of bacteria to do research with.

The genome of *M. leprae* has been completely sequenced, and this confirmed that it contains many genes that became inactivated during its evolution, and are no longer functional. But instead of losing these (like other endosymbionts have done, as exemplified in Chapters 8 and 16), all these useless genes are still present in the genome for no obvious reason. *M. leprae* has infected humans for millennia. The earliest human remains bearing convincing evidence of leprosy are 4000 years old and were discovered in India. Nevertheless, genetic evidence of the bacterial genome suggests that the disease may have originated in Africa. From comparison of mutations found in the genes of isolates from all over the world, the routes by which this disease spread could be mapped. For instance, the Silk Road, linking civilizations of China and Europe in Antiquity, has contributed to the spread of leprosy, but the disease traveled from West to East, in the opposite direction to the silk that was traded. It could also be established that the disease was introduced into the Americas by European colonists (together with many other infectious diseases). Prior to their arrival, leprosy was absent in the endogenous population. However, one observation remains unexplained. The large number of (defective) genes in *M. leprae* must have lost their function millions of years ago, a development that could only have been the result of an endosymbiotic lifestyle. But the human race is not that old, so the bacteria must have adapted to this new life in an alternative host, whereas the host jump to humans must have happened far more recently. Which host donated this unwanted disease to early humans remains a question for future researchers to tackle.

Animals, wild or domesticated, can transfer diseases to humans, whereas the reverse is true as well: humans are also a source of infection to animals. This is probably the route that *Mycobacterium bovis* took, a bacterium causing tuberculosis in cattle that is closely related to *M. tuberculosis*. It most likely originated from human *M. tuberculosis*, and made a host jump from human to cattle between 10,000 and 20,000 years ago, as estimated by genetic comparison. This time span can be narrowed down, as the host jump probably occurred after the domestication of livestock, which took place approximately 13,000 years ago. Since then, cattle infected with *M. bovis* have become a source of infection to humans. Thus, a pathogen that cattle originally got from humans evolved into a new species, and now infected cattle can again infect humans. The symptoms of infection are very similar to that of tuberculosis. Other *Mycobacterium* species exist that have adapted to animals, and all these may have originated from *M. tuberculosis*. The earliest evidence of humans suffering from tuberculosis comes from 5000-year-old human remains, as this disease can also result in lesions of the bone. It has been suggested that humans took *M. tuberculosis* out of Africa (where it may have originated) as they expanded their territory, and that all animal-specific *Mycobacterium* types have developed since. Nowadays, *M. bovis* is mostly a veterinary disease in developed countries, where eradication strategies aim to remove the disease from cattle entirely. But in developing countries, *M. bovis* can be a cause of human tuberculosis with unknown importance. Where on the globe this pathogen made its first host jump from humans to cattle remains uncertain, but it may have happened in Europe, since all current infections can be traced back to European strains. It is believed that bovine tuberculosis conquered the world from this continent.

The introduction of larger agrarian populations was another major factor that changed the scale and incidence of infectious diseases. Hunters and gatherers or herders will have suffered from infectious diseases, but it is unlikely that they suffered from epidemics. The small size of nomadic clans would not have allowed large epidemics to develop, but this changed as populations grew larger and towns or cities developed. The effect of population size on epidemics can be illustrated with data from China. Historical records of the dates and places of outbreaks that occurred in the Chinese Empire are available from the First Emperor Qin (221–206 BC) to the last Manchu dynasty, ending in 1911. The records list 488 outbreaks of disease during these centuries. For the first 300 years or so, outbreaks were rare, after which a relatively constant number of approximately 10 outbreaks per century occurred for the period 100–1100 AD. In the following centuries, the number of outbreaks grew rapidly, to 80 outbreaks per century after 1800. This pattern closely matched the population density, as far as we can tell from (incomplete) census data. The long period with a low and constant number of outbreaks matched a constant population size, but after 1100 this started to increase exponentially. The Chinese burden of epidemics grew at the same rate as the human population. We do not know what pathogens caused these epidemics, but most likely, a correlation between population size and the occurrence of epidemics must have held for other societies as well.

Infectious diseases are not only a threat or a nuisance but they also have an evolutionary role, in that they drive the selection of mutations. This is an ongoing process in all organisms, as all living things are under attack from infections. Bacteria are infected by phages, protozoa by viruses and bacteria, and higher eukaryotes (including people) can be infected by viruses, bacteria, and protozoa as well as by other eukaryotes. Infectious diseases have shaped our genes, and left their imprints on our genomes. The best examples to illustrate this come from either viral or protozoan infections, although bacteria undoubtedly played their part as well.

Malaria is very common in tropical areas, and *Plasmodium falciparum*, the protozoa causing the most severe form of this disease, multiply in red blood cells (erythrocytes) of infected patients. Some individuals are more resistant to malaria than others, and these people have a mutation in one of their genes for hemoglobin (the protein in red blood cells that transports oxygen). Humans have a double set of chromosomes in their cells, so every gene is present twice. When one of the hemoglobin genes contains this particular mutation, the other gene can still produce normal protein, and with half of all hemoglobin being normal, the carrier of the mutation is still healthy. But the effect of the mutated protein is that *P. falciparum* can no longer effectively multiply in their erythrocytes, so that these individuals are protected against malaria. This results in higher life expectancies in malaria-ridden areas, with better chances of reproduction, so that the mutation increased in the population over time. Inhabitants of countries where malaria is endemic far more often carry this mutation in their genes than human populations in other regions. There is a price to pay, though. If a baby is born with two copies of the mutated gene, it develops a serious condition called sickle-cell disease, where all red blood cells are maldeformed and malfunctioning, as they cannot produce normal hemoglobin. Without proper treatment, these patients have a low life expectancy.

A similar "enrichment" of a mutation as a result of an infectious disease may currently be ongoing during the AIDS/HIV epidemic. There are infected people who miraculously live with HIV without getting AIDS, owing to a mutation in one of their genes. Others never get infected despite multiple exposures to the virus. This mutation is more common in Northern than in Southern Europeans, and seems to be absent or very rare in Asia or Africa. If the HIV epidemic were to run its course without medical intervention, like epidemics have done in the past, it could be expected that in the surviving population the mutation would be far more common. It would also predict that AIDS would take the largest toll in those areas where the protective mutation is rare or absent. This process may be taking place during the current epidemic to some degree, notably in countries where medication is still insufficient.

It has been speculated that the mutation that provides some protection against AIDS may have been enriched in the population because of an epidemic in the past, and the plague (see Chapter 19) was proposed as a likely candidate. However, a number of observations contradict this explanation. The Black Death took the highest toll in Southern Europe, where the mutation is less common than in Northern Europe. Moreover, the plague arrived in Europe from Central Asia, whereas the mutation is absent there. As discussed in Chapter 19, a lot of uncertainties remain over the nature of the Black Death epidemic, and some scientists still question whether *Yersinia pestis* was the cause. But whatever its cause, it almost certainly did not select the mutation that currently protects a number of people against AIDS. On a positive note, the discovery of this mutation has resulted in novel medication against this terrible disease.

Infectious diseases select for mutations in individuals, but can also influence the fate of societies. Large parts of Africa were at a major disadvantage for development of agriculture, because of the presence of the tsetse fly. This insect prefers humidity, shade, and a temperature between 20 and 30°C, and lives in a large area approximately between the Sahara and the Kalahari Desert. Importantly, the bloodsucking flies kill domestic animals because they spread *Trypanosoma*, the cause of sleeping sickness. As a consequence, it was nearly impossible to herd animals in infested areas. The lack of animals to pull the plows also hampered the development of agriculture. Living was difficult in areas where the tsetse fly lived, which led to diseases in animals and people alike. Even today, the effect of the tsetse fly and its deadly disease coincides with the occurrence of lactose intolerance in humans who are descendants of farmers instead of herders. Sadly, after the disease was on the wane in the second half of the last century, trypanosomiasis is back to alarming levels. In Angola, for instance, the number of cases had decreased to close to zero in 1975 but increased again to over 12,000 in 2001. The tsetse fly and the disease it transmits may be one of the reasons why the African continent remained underdeveloped compared to other continents.

The future may bring new patterns of infectious disease, owing to climate change and the expected human migration patterns this may cause. Changes in local climate and land use will have an effect on disease burden, but what can we expect? Several researchers have tried to predict how climate change will change infectious diseases, but such predictions are mostly speculation. Observations from the past are no guarantee for future developments. Many factors influence the rise and fall of infectious diseases, and to predict these based on expected climate changes is near to impossible, especially since the extent and speed by which the climate might heat up (or, depending on the location, regionally cool down) is again uncertain. Depending on location, either heavy rainfall or droughts can become more common. The areas where insect vectors are endemic might change, whereby their living quarters may not always expand; some insects, and the diseases they carry, may shift location instead, so that a disease may be emerging in one place but disappear from another. Thirty years ago, nobody would have predicted that HIV/AIDS would kill millions of people, and that epidemic was not caused by climate change. With this degree of uncertainty, trying to predict the future of infectious diseases is mostly science fiction. All we can do is to remain on guard. Microbes will continue to surprise mankind, as we humans are only humble members of a global ecosystem in which microbes are key players, whether we like it or not.

23

Big Questions on Small Subjects

The subjects treated in this book were selected to explain some of the fundamental characteristics of bacteria, to explore their diversity, and to illustrate some of the scientific insights that have resulted from years of hard work by numerous microbiologists. Since scientists keep generating novel data, it is valid to ask what we can expect from their work during the next decade. Which subjects are likely to fascinate researchers in the future? Where can we expect, or hope for, significant breakthroughs? Crystal ball predictions are bound to be inaccurate, but a few trends can be recognized from current work. This chapter sums up some of the questions that occupy the minds of present bacteriologists that may well lead to future discoveries.

Traditionally, the main focus of microbiological research has been infectious diseases and how to treat or prevent these, and it is expected this will remain a focus of attention in the future. A lot of work is wanted to develop novel antibiotic drugs, which we urgently need to fight infections caused by bacteria that have become resistant to existing ones. Research of drug design is traditionally in the hands of the pharmaceutical industry, but the identification of novel critical processes taking place in bacteria that can be targeted by drugs results from fundamental research, which is mainly carried out at academic institutions. The collaboration between fundamental academic research and the pharmaceutical industry can sometimes be problematic, and antibiotics do not form the high profit blockbusters that the industry has lately been most interested in, but there are trends that this may change. The pressure to produce new antibiotic compounds is now stronger than

Bacteria: The Benign, the Bad, and the Beautiful, First Edition. Trudy M. Wassenaar.

ever. Instead of the minute chemical changes that the pharmaceutical industry is mostly interested in (such changes can result in new patents), a fundamentally different approach is needed. Maybe we should turn away from broad-spectrum antibiotics, and design more specific drugs, which, combined with faster and more accurate microbial diagnosis, can eliminate a specific pathogen without harming residual microflora. This would require a change in prescription habits as well as a change in patient's expectation: an antibiotic should then only be given after experimental confirmation of the infective agent.

The vaccine industry faces a different problem. Vaccination has been such a success in developed countries that unexpectedly social resignation emerged. The diseases these vaccines prevent have become so uncommon that the public no longer appreciates the risk that these infections pose on human health. Instead, the side effects and complications of vaccination are being perceived as a serious "danger" by some. Side effects of vaccinations are in most cases mild, but in rare instances, serious complications can occur. These harms are still far less numerous and often less serious than the harm caused by the disease in unvaccinated individuals. When a large proportion of the human population chooses to remain unvaccinated, however, the disease will become more common and so will be their complications that can result in permanent disabilities or death. The subtlety of this risk evaluation is lost under an avalanche of disinformation, spread by mouth and particularly via the Internet.

In 1998, the United Kingdom saw a drop of 25% in vaccination acceptance for the MMR combination vaccine (against measles, mumps, and rubella, all three viral infections) after rumors were spread that this could be linked to autism, a suggestion that was subsequently proven wrong. Possibly because of the decreased vaccination rate, a mumps epidemic occurred between 2003 and 2006 involving over 100,000 patients. These were mostly unvaccinated adolescents (in whom the disease can cause male infertility), but the mumps virus may have sheltered in unvaccinated babies, thus intensifying the epidemic. A recent effort to vaccinate adolescent girls against human papilloma virus, in an attempt to prevent cervical cancer later in life, failed miserably in the Netherlands. This country is known for a high public acceptance of vaccination, but the introduction of this new vaccine went anything but smoothly, partially as a result of misinformation and slander spread via the Internet. Well-informed people weigh the (small) risk of cancer against the discomfort of an injection. Immunologists will have to inform and clarify the public more convincingly why novel vaccines are necessary and how to balance the advantages and risks, in order to gain public acceptance. Without that, vaccination programs are bound to fail.

Vaccine development of the future should especially concentrate on fighting infections in developing countries, where the gains in terms of public health can be tremendous. The vaccines developed in the past are not very practical in conditions where electricity is absent so refrigerators do not work, sanitation is insufficient, and clean needles are not available. Novel means to administer vaccines, for instance, by oral application or nose sprays instead of injections are currently being developed.

The idea to administer antibiotics, vaccines, or therapeutic drugs by means of polymer-coated microparticles is not novel, but it is gaining momentum as the technology to produce sprays of bioactive molecules is improving. It is hoped that this will eventually lead to cheaper vaccines that are more easily transportable and can be administered under conditions that apply in those areas where vaccination is most wanted.

Meanwhile, the developed countries face novel diseases that grow to epidemic proportions and are related to a Western lifestyle. Obesity, type II diabetes, heart disease, and allergies are on the rise, and fingers are pointed toward bacteria as suspects. These diseases are most likely multifactorial, but if bacteria are one of the causes, this may provide a target for treatment and prevention. The research concentrates not only on the definition of "obese" and "lean" gut microflora (see Chapter 20) but also on the role of the host's immune system, as it is recognized that these diseases all have an immunological component. It has been well studied how pathogens are attacked and removed by the immune system and how some pathogens withstand these defenses. However, how the human body responds to our commensal bacteria is far less clear, although we do know that there is continuous interaction between our immune cells and our bacteria on a daily basis. There may be key events in this interaction that, when things are not optimal, eventually lead to obesity and diabetes, allergies, or cardiovascular disease. Novel insights in this field are expected in the next few years, as a lot of research resources are diverted to this important and relatively new field of host–microbe interactions. The expected insights may well have implications to our better understanding of "Western" diseases, potentially leading to better treatment or more effective prevention strategies.

Commensal gut bacteria may even influence our brain. It is long known that the central nervous system regulates and thus influences gut functionality, but more recently, it was discovered that there is an interaction in the other direction as well, whereby immune cells are again the key. Stress can affect the gut microbiota, as was directly demonstrated in mice; studies have also suggested that the presence of gut bacteria influences the stress responses in these animals, coinciding with the development of their immune cells. Although it is difficult to extrapolate these results to humans, it is likely that our gut microflora and brain development are somehow related. Many gastrointestinal disorders occur simultaneously with psychiatric conditions. It is currently studied if and how the gut microflora can be influenced by food supplements, and we can expect to hear more about this subject in due time. Meanwhile, lawyers are already considering whether they can blame the gut bacteria of their clients in defense of their misdoings.

The interaction between our benign bacteria and our body will remain a hot topic, with far-reaching implications, as the examples above exemplify. The immune system seems to have a "hand" in everything that is going on in our body, and one of the first triggers that set the immune system of a newborn going is the microflora it encounters in the developing gut. For the rest of our life, immune cells and bacteria communicate, mostly (but possibly not exclusively) in our gut.

It seems like the question to be addressed has changed from "whether" to "how" bacteria can improve our health.

It is hard to imagine that bacteria can serve as a model to study complex processes in humans, but it so happens that many processes taking place inside bacteria are not so different from those of our own cells. One such process is our circadian clock, which dictates we sleep at night and are awake during the day. If we reverse our sleeping pattern and take night shifts, or change time zones after an east- or west-bound flight, our clock is out of phase and needs time to adjust. Our cells remember when it is supposed to be night or day, and reluctantly give in to a new time schedule. This was considered a complicated and intricate process, until a Japanese research group was able to mimic a circadian clock in a test tube. They only had to mix three proteins that they had isolated from *Synechococcus elongatus* to mimic the cyclic activity that regulates the circadian clock of these Cyanobacteria. For these photosynthetic organisms, it is important to anticipate when it will be light, and their cellular processes oscillate during light and dark phases. Surprisingly, the discovery of how this works in Cyanobacteria can explain some of the key features of the human circadian clock, too. This illustrates how bacteria can serve as a model to study complex processes in higher organisms.

Many biologists have been fascinated by the question of how life started, and microbiologists are no exception. What are the requirements, and how did life evolve into its present forms? As we have seen in Chapter 6, microbes (bacteria, endosymbionts) and viruses can exist at the border of life and nonlife. This will remain an interesting aspect of microbiology where it crosses the path of philosophy. The next chapter will introduce synthetic life as an appetizer for future expectations.

Another motive for research is curiosity, the urge to discover. What is living out there? Which microbes live in the ocean, inside and on plants and animals, in the soil? What can we learn from the undiscovered biological richness in these niches? We can be certain that surprises will await us. One promising application of this explorative research is that it may eventually lead to new drugs. Bacterial products are so diverse that, inevitably, some of these have practical applications for us. Especially marine microbial life may have good things in store, and it is expected that in the next few years, bioprospecting of marine bacteria becomes feasible.

A further drive to do research is the wish to understand how things work. How do genes function? How were they formed, which organisms can do what, and how, and when? Bacteria are key players in larger ecosystems, and the first metagenomic sequencing projects have identified a wealth of genetic richness that we have only started to investigate. The diversity of bacteria was emphasized time and again on these pages, but it turns out that this is even greater than the known existence of phyla, genera, and species that have traditionally been studied. As techniques that allow analysis of single cells were developed, it was discovered that even within a colony, cells that were all derived from one single progenitor are not equal. Microdiversity exists in the way they express their genes and respond to their environment. The importance of this cell-to-cell heterogeneity for the complete population is not yet known, and this is an exciting field of research, which may result in insights that can also improve our understanding of higher, multicellular organisms.

Applied scientists have different drives for their research. They want to solve problems. The field of bioremediation is an important and rapidly growing discipline within microbiology. The problem of humans increasing the atmospheric CO_2 concentrations, by burning fossil fuels, was extensively treated in Chapter 14. This is not the only gas we have muddled with; however, that human activity has also largely increased the nitrogen levels in surface waters is less generally known. Over the past century, the use of fertilizers, needed to satisfy a growing global demand for food, has drastically disrupted the nitrogen cycle. Too much nitrate that eventually reached rivers, lakes, and coastal waters is now leading to algal blooms. The excess nitrogen will eventually be released back into the atmosphere thanks to denitrifying bacteria, but this can take centuries. Bioremediation to restore the nitrogen cycle is an example of applied microbiology that attempts to restore some of the effects our society has had on our planet. Other applications of bioremediation with the help of bacteria that are under study include the clean up of industrial waste, deposits of toxic heavy metals, and oil spills.

It was already mentioned in Chapter 14 how photosynthetic bacteria and algae may eventually provide a climate-friendly alternative to petroleum. Imagine the advantages. Algae and Cyanobacteria can grow in saline water, of which our planet has plenty (although nonarable land has also been mentioned as a place to farm bacteria), and their production would not compete with food production. They would use carbon dioxide from the air and transform this, with the use of energy from sunlight, to sugar, starch, and oil; their light-to-biomass conversion is as efficient as that of plants, but they occupy far less space. After harvesting their high energy produce, the remains can be pressed and heated into a charcoal-like biochar, which can be used either as a coal alternative, buried to capture atmospheric CO_2, or as a soil fertilizer. However, we are not there yet. Not one algal biofuel system has achieved economic viability yet, and future research has to solve a number of serious difficulties. Nonarable land would often not be suitable to grow bacteria or algae, which need nutrients, water, and ambient temperatures just like plants. The efficiency of converting light into biomass has to increase approximately threefold to be economical. Scientists are studying Rubisco, the key enzyme needed for carbon fixation, to see if its poor efficiency can be improved. For a start-up, species could be chosen that produce, apart from fuel, valuable pigments or drugs, which will increase the profitability. Whether this idea will become economically feasible depends on many factors, but microbiologists will have a major influence to increase a chance of success.

When less petroleum is being used, products derived from this oil will have to be replaced by alternatives. All plastics are traditionally produced from oil hydrocarbons. Plastics are widely used in households and in medical and industrial applications, but the downside of these materials is that they are degraded very slowly and burning of plastics produces toxic compounds. A suitable alternative may be plastics based on naturally occurring polymers, which are produced independently of oil and are biodegradable. The first natural polyester was already discovered in the 1920s by Maurice Lemoigne (1883–1967) from France, who observed the small granules inside the cells of *Bacillus megaterium*. This polyester is a member of the polyhydroxyalkanoates, abbreviated as PHAs, which are made by a large number of bacteria and even by some plants. PHA polymers are biodegradable so that they will not accumulate in the environment, and they are immunologically inert, which means they could be used in medical applications. Some bacteria produce the polymers following heat stress or starvation, but a few, such as *Azotobacter vinlandii*, produce PHAs all the time, which makes them more interesting for large-scale production. At present, production on an industrial scale is not yet feasible, and the polymers are not yet widely used, but these bioplastics have potential for the future.

One way to improve industrial application of bacteria is to change their genetic properties so that their products suit our demands better, which will be treated in the next chapter. The big and small questions that are currently being investigated indicate that microbiological research is thriving, and novel findings will probably keep hitting the headlines. These headlines can sometimes be a bit too enthusiastic. It should be remembered that scientific "breakthroughs" are relatively uncommon. Despite the flashy news bites and exaggerated press releases that seem to be inevitable in our information-hungry society, most scientific insights develop from small but steady increments of knowledge. Scientists could better refrain from overoptimistic promises that their discoveries might hold. As these chapters have illustrated, the bare facts are fascinating enough.

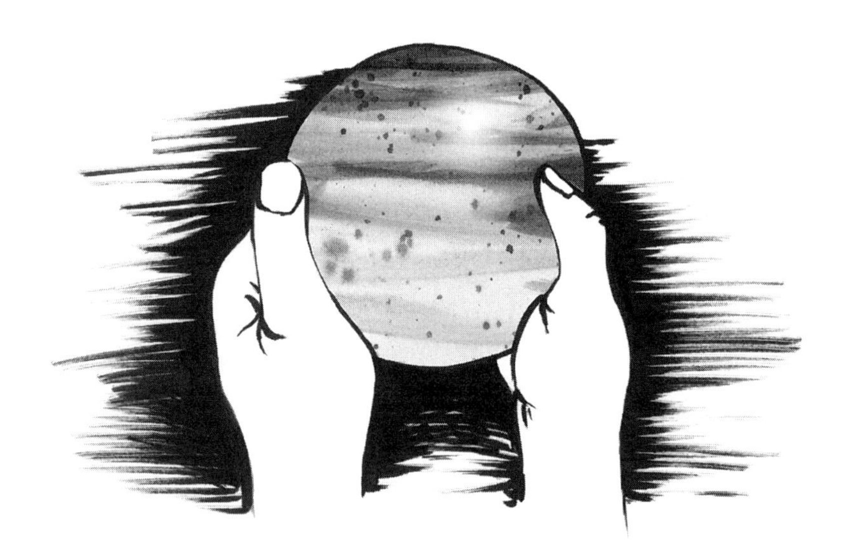

24

Synthetic Biology

Consumers, politicians, and legislators alike struggle with the acceptability of genetically modified organisms. The definition, according to the Encyclopaedia Britannica, of genetically modified organisms is "organisms whose genomes have been engineered in the laboratory in order to favor the expression of desired physiological traits or the production of desired biological products." For genetically modified plants or animals intended to be used as food or feed, a distinction is often made between genetic changes that could have occurred naturally (mostly those that were or could have been obtained by classical breeding) and those that would not naturally occur; especially, the latter will have to go through a number of safety assessments before they can be released in the environment or enter the food chain.

In Chapter 11, it was explained that the introduction of changes into the genetic material of bacteria is a common procedure. As a result of human interventions, genes are frequently being inactivated, deleted, or inserted into bacterial genomes; during this process, these genes can be altered or moved around or their copy number can be changed. In most cases, it is hard to say if the changes made could have occurred naturally or not, as bacterial DNA is subject to natural changes on a large scale in Nature. The limitations of these processes are not known, so we cannot really separate the "naturally possible" from "unnatural" events. If bacterial genomes are so vulnerable to natural change, why do not we observe this in the laboratory?

Bacteria: The Benign, the Bad, and the Beautiful, First Edition. Trudy M. Wassenaar.

When bacteria are grown in the laboratory, their genomes remain pretty stable, but the way we pamper our research organisms cannot be compared to the harsh conditions they meet in the environment. Bacteria are typically cultured in the laboratory without the presence of other species that would compete for food and might produce toxins to claim that food exclusively for themselves. Cohabiting bacteria can be infected with plasmids, which can actively or passively be transmitted to other cells, mostly within but sometimes between species. But a bacteriologist will take all possible measures to avoid growth of other bacteria that would contaminate their cultures. Viruses are absent from laboratory cultures but regularly attack bacteria in their natural environment, frequently donating or stealing genes as they infect strains one after another. Viral attack is extremely common in the bacterial world. There are so many virus particles present in the ocean, constantly infecting and killing bacteria and algae, and less frequently plants and animals, that they cause a turnover of approximately 20% of all marine biomass every couple of days. Thinking of the attacks bacteria undergo in real life, one can only pity them for their struggle to survive in the environment.

Viruses and plasmids often carry mobile elements: selfish parasitic DNA such as transposons that insert into chromosomes and relocate themselves frequently, knocking around genes as they do, but all of that is prevented in a typical laboratory culture. We do not stress our bacteria by heat or cold and protect them from sunlight or radiation or mutagenic chemicals. In short, we do everything to keep them growing as they do, without providing them with the needs, or means, to change their DNA and without selecting for any change. This is why their genes and genomes appear so stable. Another faction to consider is time: growing a bacterial species for 25 years in the laboratory (which has been done, and mutations have been observed to accumulate during this period) is a long time, in terms of scientific interest and funding, but it is only a waft on an evolutionary time scale.

Some of the natural perils for bacterial DNA are actually used by geneticists: transposons can come handy if you want to insert novel genes into a bacterial chromosome, and plasmids are frequently used "vectors" to put DNA from one cell into the next. (As an aside, the "peril" of natural DNA-changing phenomena is relative, as they are important drivers of evolution. Without viruses, plasmids, selfish mobile elements, and radiation or chemical damage, bacterial genomes would change less rapidly, meaning their populations would be less adaptable to novel conditions and environments. Viruses and other selfish DNA are main drivers of evolution, although they are not alive.)

As explained in Chapter 11, when necessary genetic modification of bacteria is performed, safety measures are in place to protect the researchers and the environment: safety protocols are there to either prevent release of potentially dangerous bacteria into the environment or prevent their survival, in case they would escape. These measures depend on the assessed risk of the mutants that are being produced. In some cases, these measures can go far. Some experiments are done in organisms that depend on an artificial building block for their DNA, which they cannot make themselves. Such mutants can grow only in the laboratory and as long as their necessary DNA nucleotides are supplied in their medium; since these nucleotides do not occur naturally, there is no way these bugs could live outside a laboratory. These experiments also provide insights into the limits of what is biologically possible: not all that is possible is actually happening in Nature, and by pushing the limits, fundamental insights are obtained as to how a cell works. It has been shown that special amino acids, which do not occur in Nature, can nevertheless be incorporated into the proteins of bacteria, as long as they are provided with the necessary details of how and when to use the alternative amino acid. Scientists have even been able to even alter the typical three-nucleotide code for every amino acid into a four-nucleotide code.

The applications of genetically engineered bacteria are plentiful. Since the production of pharmaceuticals is an important application, a historical example from this field is given here. One of the first breakthroughs in this technology was the construction of *E. coli* cells producing insulin. Insulin is a protein (it is counted among the hormones) that regulates the level of blood sugar; it is normally produced by the pancreas when blood sugar levels rise because of sugar intake. As a result of the insulin released into the blood, liver cells start storing away this sugar to prevent levels reaching dangerously high concentrations. The stored sugar is subsequently released when blood sugar levels drop. Diabetic patients do not produce (enough) insulin (often as a result of an autoimmune disorder in which insulin-producing cells are under attack of their own antibodies), or their cells have become insensitive to the levels their pancreas can produce. Many of these patients need regular insulin shots to regulate their blood sugar levels. Traditionally, medicinally used insulin was isolated from animal pancreas, but its purification was expensive, and there was a risk that an animal virus could contaminate the product. After the discovery of insulin and its function (which earned Frederick Grant Banting and John J.R. Macleod from Canada the Nobel Prize of Medicine in 1924), it was chemically produced in 1966, the first hormone to be synthesized in a laboratory, but its chemical production was not practical on an industrial scale. In 1982, the human insulin gene was cloned into *E. coli*, which, as a result, produced biologically active insulin. This recombinant insulin (the term means it was produced by DNA recombination, but the product is indistinguishable from natural insulin) is still in use today.

Once the insulin gene was properly inserted into the *E. coli* genome, scientists were able to mutate it and test if the resulting altered protein would function differently from its natural, or "wild-type," version. In some cases it did, and its altered behavior was beneficial, so that now recombinant insulin is available that works faster than natural insulin or slower but more steadily, depending on how the protein gene was changed. These altered insulin drugs serve the needs of patients better than the natural hormone, given the fact that medication is administered at time intervals and doses that are not as fine-tuned as can be done by a healthy pancreas. The example illustrates the possibilities of the recombinant DNA technology: genes can be altered so that they produce proteins "designed" to have better properties. Sometimes the desired changes are indeed designed, but more often, useful mutants are found by trial and error. The process of mutation and phenotypic selection by trial and error has become an art by itself, as it has been optimized to allow high-throughput tests.

It sounds simple enough: add a gene to *E. coli* and let it produce the protein of choice. In practice, there are a number of difficulties to overcome before this works. Most genes of eukaryotes are cut up in shorter sections, interrupted by DNA that does not contain any protein-coding information. The segments that code for pieces of the protein are called *exons*, and the intermittent, noncoding sections are called *introns*. The eukaryotic cell produces a messenger RNA molecule of this mixed lot, which is then cut and glued to remove the intron segments, a process called *RNA splicing*. Bacterial genes are hardly ever interrupted this way, and bacteria lack the machinery for RNA splicing: they cannot remove introns from messenger RNA. For a eukaryotic gene to be correctly read by bacteria, introns have to be removed from the gene first. Moreover, many eukaryotic proteins require that specific sugar molecules are attached to them, in order to function properly: the protein has to be *glycosylated* by enzymes. Again, protein glycosylation is less common in bacteria, and even if they glycosylate some of their own proteins (flagellin is often glycosylated), their glycosylation enzymes would not recognize the correct positions on the naked foreign protein they are forced to produce.

Some restrictions to the recombinant technique become apparent only when experiments leave the drawing table and enter the reagent glass. The protein we force bacteria to produce should not be toxic to them, but toxicity cannot always be predicted with accuracy. Likewise, the bacteria should not produce proteases (enzymes that break down proteins) that attack the harvest. Finally, and crucially, the protein produced should leave the bacterial cell, so that it does not build up inside the cell (with likely toxic effects) but is released in the liquid medium for easy recovery.

Protein secretion is a complex process in a cell. Proteins cannot pass the membrane or membranes that surround the cell without being actively transported. Specific secretion systems have evolved to sluice proteins out of cells, of which the TTSS has already been introduced (see Chapter 5). There are at least six different types of secretion systems produced by bacteria, and each can transport only particular proteins across the membrane. Most of these systems secrete proteins into the medium, whereas the TTSS is special in that it ejects the proteins into target cells. Typically, a secretion system recognizes which protein has to leave the cell because this protein carries a label: a *secretion signal,* usually a short segment of amino acids in a particular order. This label is recognized by the specific secretion system in the membrane of the cell. The artificially introduced eukaryotic protein will most likely not contain such a bacterial secretion label, so that has to be added to its gene by genetic engineering, or else the protein would get stuck inside the bacterial cell. Such secretion signals are frequently cut off from the protein as it exits the cell, so this artificial addition would not be found on the protein once it is secreted into the medium, from which it can be recovered and purified.

The production of recombinant protein by microorganisms that do not naturally produce these can be regarded as "synthetic biology." However, that term is not restricted to the odd gene added to an existing bacterium. *Synthetic biology* aims to design and construct complex artificial biological systems with the use of existing bacteria and a set of well-characterized genes. In synthetic biology, scientists use standard biological parts, much like an engineer designing an electronic or mechanical system. A number of genes are typically needed to produce, modify, and secrete a desired product. In the example of insulin, introduction of the gene coding this protein into *E. coli* was sufficient for its production. But more typically, the aim is to produce a biomolecule that is produced by enzymes, rather than being the direct product of a single gene. For instance, to produce a pigment that isn't a protein, so that there is no gene that codes for it directly, a complete metabolic pathway may have to be introduced, to insert all genes coding for all enzymes that eventually produce the foreign pigment. One of the challenges for a synthetic biologist is to optimize the pathways by designing clever combinations of existing genes that do not naturally collaborate in a biological production line. This technology is applied to modify and optimize bacteria in order to produce biofuel, and some pilot experiments are promising, although whether this will work on a large scale still has to be put to the test.

Synthetic biology has created DNA that incorporates building blocks that do not occur naturally. Scientists have even designed an extra base pair, in addition to the two that are found in all natural DNA (the A-T and the C-G base pair), so that this synthetic DNA can carry information with six rather than four different nucleotides. It shows that there is no fundamental reason why DNA exists with four bases only; an intelligent designer could have added some more variation to the scene.

The bacterial sensors described in Chapter 21 as part of two-component signal transduction systems are another source of human inspiration. The sensor systems of bacteria could become the "eyes" and the response regulators the "brain" of living, self-replicating computers that can interact with the physical world. Although this sounds like a scenario for a science fiction film, it is more science than fiction: there already exist "how-to" guides that explain the practicalities of building living computers. In all fairness, however, their applications are, as yet, still mostly hypothetical.

Is anybody looking into the risks of such breakthroughs? Should we pause research and start a discussion like we did nearly four decades ago, as described in Chapter 11? It is unlikely that scientists can now be halted, and novel techniques are currently received with more acceptance than before, although there are always critical minds, and these rightly express their concerns. This keeps us aware of what we are doing. It seems, however, that the present-day breakthroughs are better welcomed than those that started the whole new field of synthetic biology, the simple experiments that produced the first genetically modified bacteria back in the 1970s.

With advancing technical developments, eventually attempts were made to create artificial life from nonliving material. We can produce DNA from nonliving chemicals, and by tedious chemistry, we can even produce DNA that is long enough to resemble the genomes of simple bacteria. However, as discussed in Chapter 6, DNA in a test tube is not alive and would just stay put. It needs a cell around it in order to function, and the cell needs DNA in order to function. Inside a cell, DNA is not alone; it is covered by proteins that bind to it and curl and twist it: only then can the cell "use" DNA and "read" its information. Nevertheless, scientists have produced artificial life in the laboratory, starting with DNA that was chemically produced and bringing it to life. They produced artificial DNA that was nearly an exact copy of that of a natural bacterium, *Mycoplasma mycoides*, so although it had been produced chemically and not biologically, it was identical to what Nature normally makes without us paying much attention.

The vicious circle of DNA needing a cell and a cell needing DNA was broken by a little cheating. After long fragments of the artificial *Mycoplasma* bacterial genome were produced in a test tube, they were put into yeast cells by means of a plasmid. These independent DNA molecules will be reproduced by the yeast cell without much trouble. Once inside, the *Mycoplasma* fragments found each other while yeast enzymes, doing the DNA repair job they are supposed to do, linked them together into a complete genome existing inside a yeast cell. The DNA was then turned and twisted to condense it in the way genomes typically are found inside cells.

Yeast is a eukaryote, that is, most of its DNA is packed in a nucleus. The yeast cell could not use the DNA it was given artificially, as this existed in its cytoplasm and not inside its nucleus. So the complete *Mycoplasma* genome could exist inside a yeast cell without interfering with the yeast's metabolism. The artificial genome was then isolated and shot into a "ghost cell": a real, living bacterium that had its DNA completely removed but still possessed an intact membrane and most of its proteins. On receipt, the proteins started binding to the naked DNA and made it functional so that it took over all cellular processes. The cells started dividing, and after a few generations, all cells contained copies of the new DNA with all its genes, and contained proteins that were produced from these genes.

This technical masterpiece, performed in the laboratory of J.C. Venter (the same that sequenced the first human genome) has far-reaching implications. Is it safe to produce life? Is it ethical? What can we do with it? Can it be misused? As with every technical breakthrough, there are supporters and admirers, but there are also sceptics and the project received some serious opposition. The supporters hailed the possibilities: we could produce artificial bacteria that can fix carbon dioxide from the air to produce oil, which we could burn without affecting our climate. We could design and produce bacteria to fight off diseases. Combine this with the toolboxes of synthetic biologists, and everything might be possible. In all fairness, however, such desired bacteria can also be produced by modifying existing ones, instead of starting with abiotic chemicals, as the sceptics pointed out.

The objections concentrate on potential dangers that also linger in this novel technology. Venter's bacteria were as safe as their natural counterparts, as they were nearly identical copies, so there was no danger of upsetting an ecosystem or starting an epidemic. But the technique could result in such unwanted side effects, for example, by producing bacteria with as yet unknown properties. If creating artificial life (which is far from simple and extremely expensive) were to be misused, a lot of harm could be done; for instance, if rogue scientists were to create deadly bacteria deliberately (but then again, there are a wide variety of deadly bacteria available naturally, so why bother producing a novel one from scratch?) The question whether it is ethical to create or manipulate life in a test tube frequently addresses only potential experiments with animals or even human clones, whereas bacteria are often ignored, but the recent achievements in this field have pointed out the risks and possibilities that are at stake. The production of artificial bacteria is still very distinct from artificially creating animals or bringing extinct species back to life. Nevertheless, artificial life now exists on earth, and it was created by man. One just cannot easily tell the difference between the bugs that cost 40 million dollars to produce and their natural siblings.

25

Bacteria, the Earth, and Beyond

This book started with a global view of our planet, and that perspective will finish this final chapter as well. As we have seen in Chapters 3, 13, and 18, the major gases of the earth's atmosphere, nitrogen, oxygen, and carbon dioxide, are all used and produced by bacteria. With every discovery of the abilities of bacteria, it was speculated that we could manipulate these magical properties and use them to our own good. As stated in the previous chapter, scientists study their pet subjects to satisfy a need to understand things, to be able to explain how things work, and maybe even to know why things work as they do. But people also want to manipulate. We do not only observe and describe our surroundings, we try to change them all the time. A constant stream of ideas is proposed how we could manipulate bacteria to change things for the better, or so it is believed. Bacteria could be used to improve our health and combat obesity and allergies. They could improve our environment, clean up our waste, or produce biofuel for us. They might be used to produce rain to combat draughts, or remove arsenic from naturally poisoned, man-made water wells that kill so many people in Bangladesh. Bacteria could, maybe, restore the carbon dioxide levels to preindustrial levels and abrogate the much-dreaded global warming. Politicians or economists rarely think or talk about bacteria, but the effects that these little cells can have on our planet, with or without some manipulation, is of global importance.

One of the newer insights in microbial ecology is that bacteria communicate and interact with each other. Not only with cells of their own kind but also with other species that share their environment. Most microbiologists still study bacteria in

Bacteria: The Benign, the Bad, and the Beautiful, First Edition. Trudy M. Wassenaar.

the form of pure cultures, where all cells are made up of the same species. But that is not how most bacteria live in the real world. Pure cultures are a rarity in Nature, and mixed cultures are the norm. Adding more cultures increases complexity, and many bacteriologists will whimper at the idea of having to work with complex, unpredictable communities. However, this may be a necessary next step if we want to understand the true lives that bacteria live. Science evolves, and microbiology evolves just as fast as other fields. Who would have thought, 30 years ago, that we can sequence the complete genome of a bacterium within a day, or that bacteria live in biofilms like city-dwellers, or that they can produce injection needles?

Early in my career, I met a respected microbiologist who was a specialist in metal detoxification proteins. He could talk for hours (which he did) about the proteins that remove toxic heavy metals from the cells and release them into the environment in a less toxic form. In his long carrier (he was close to retirement) he had collected a phenomenal knowledge on the subject. Eventually, his monologue halted and he asked me, more out of politeness than with genuine interest, what subject I was working on. I replied that I was studying *Campylobacter jejuni*, a Gram-negative Proteobacterium that is a frequent cause of diarrhea (despite my fascination for this beautifully spiral bacterium, it has not played any part in this book). His face could not hide his disappointment and he gave me some earnest advice. "Change to *E. coli*, my dear," he said. "We know so much about this organism already, it is the best-studied species of all microbiology, and when you truly understand *E. coli*, you understand all bacteria."

Obviously, I did not agree. This book illustrates that most bacteria are anything but similar to each other. *E. coli* only played a minor role in some of the chapters, which were chosen to explore the vast diversity of the microbial world. How can one investigate the true world of bacteria by concentrating on one species only? His view can be excused, though. Most textbooks on microbiology from which students learn the subject are heavily biased toward *E. coli*, of which, admittedly, we know more than of any other bacterial species. But ignoring all the others would be a big mistake. The diversity of the bacterial world is divided into bacterial phyla, and every phylum and its members is equally fascinating. Quite a few phyla have been mentioned in this book.

E. coli is part of the Proteobacteria, and this phylum has been mentioned on many occasions, as it contains many well-studied genera and species. Most Gram-negative pathogens are Proteobacteria. The Proteobacteria phylum is so big that it has been subdivided into five subphyla, named alpha, beta, and so on to epsilon. *E. coli* belongs to the gamma-Proteobacteria, whereas *C. jejuni* (as well as *Helicobacter pylori*, the stomach bug that was used to separate human lineages) belong to the epsilon-Proteobacteria. *Rhodospirillum*, and those other bacteria we met in Chapter 13 whose name starts with "Rhodo" that belong to the purple bacteria and perform photosynthesis, are alpha-Proteobacteria. An example of a beta-Proteobacterium is *Neisseria gonorrhoeae*, which causes the sexually transmitted disease gonorrhea. *Bdellovibrio bacteriovorus*, the cruel one that eats other bacteria from the inside out (Chapter 16), is a delta-Proteobacterium. Many of the endosymbionts living in insects that have lost a large fraction of their genomes are Proteobacteria. We have seen that members of this phylum can deal with oil spills, for example, *Alcanivorax* species. *Pseudomonas* species are amongst those that slowly destroy pieces of art, and the remains of Romans killed by the Vesuvius eruption still carried evidence of *Brucella* infections. All these genera belong to the Proteobacteria.

The other phylum that received a lot of attention in this book is that of the Cyanobacteria, because they are so versatile. They can fix both nitrogen and carbon dioxide gas, some can form multicellular organisms and are motile because they can glide, and some produce toxins that are a problem when they rapidly multiply and form blooms. Others live as symbionts to plants, and yet others produce specialized cell forms as if they are multicellular organisms. Cyanobacteria (or bacteria like them) might have been amongst the earliest living forms on earth, and they have been the ancestors of photosynthetic plant chloroplasts. A book about bacteria would be incomplete without paying attention to the Cyanobacteria.

The Gram-positive Firmicutes were also mentioned on various occasions. They are major players in our gut and colonizers of our skin. The antibiotic resistance of *Staphylococcus aureus*, in particular the dreaded MRSA, causes doctors a headache and their patients much suffering. The most potent toxins known are produced by *Clostridium* species, while *Bacillus* species produce spores, which can persist for decades. Our cultural heritage is threatened by a variety of bacteria that are often Firmicutes.

The phylum of Actinobacteria is frequently represented as part of the human microflora, with *Bifidobacterium* as the best-studied genus. Some, but certainly not all Actinobacteria have a high G-C content, for instance, *Mycobacterium* species, or *Propionibacterium* species that produce smelly armpits. *Streptomyces* species produce many compounds that are useful, such as antibiotics and pigments. They make sweeter syrup from natural but non-sweet sugars, but Actinobacteria are not always helpful: they can also destroy paintings or frescoes. *Mycobacterium tuberculosis* has terminated the lives of many, including a number of famous artists.

A few other phyla were encountered on these pages. Members of the Gram-negative Bacteroidetes phylum are numerous in the human gut, and other species of this phylum were described to degrade cyclic hydrocarbons in oil. The Mollicutes, which are neither Gram-positive nor negative, were represented by *Mycoplasma*, which has been produced as an artificial form of life. The Chloroflexi and Chlorobi phyla were mentioned in Chapter 13 as examples of bacteria that produce bacteriochlorophyll for photosynthesis. *Borrelia burgdorferi*, the causative agent of Lyme disease, was presented as an example of a Spirochete. Another member of this phylum is *Treponema pallidum*, which causes syphilis, a disease studied by Nobel laureate Paul Ehrlich and a killer of quite a few famous composers. Bacteria belonging to the Fusobacteria phylum can start a biofilm in the mouth, and *Deinococcus* bacteria (of the phylum Deinococcus-Thermus) are known for their resilience after radiation. *Thermus aquaticus*, the heat-loving bacteria whose DNA polymerase is used in PCR, is a member of the same phylum. The phylum Planctomycetes was mentioned as the Anamox bacteria were presented, which are slow growers but effective nitrogen removers living in the ocean. A few record holders have been discussed, including *Solibacter usitatus*, a soil bacterium with an extremely large genome. It belongs to the phylum of Acidobacteria (not to be confused with the Actinobacteria phylum) which is mostly represented by soil bacteria. Still, the collection is incomplete: There are at least 10 other eubacterial phyla that have not even been mentioned, but those will be left to themselves.

The archaea were not ignored, although they never seem to receive the attention they deserve. Archaea are not only masters of extremes but are also abundant in the ocean. The four phyla that were mentioned were the Crenarchaeota (often thermophiles), Korarchaeota (which we presently cannot even culture), Euryarchaeota (mostly extremophiles), and Nanoarchaeota, the smallest prokaryotes that we know. The other phyla of the archaea sometimes contain a single species that is so weird it would not fit in anything but its own phyla, and these phyla may not even have a proper name but are designated with letters or numbers only.

Despite an effort to cover as much ground as possible, many interesting species have not featured in this book. It is impossible to aim for completeness, and mentioning more phyla and species would be boring when not put in context. Since this planet has been covered with bacteria for so long, they have developed into an array of species that is impossible to grasp, and sometimes even impossible to define, which can make the life of a researcher very difficult. Instead of a D for dirt, disease or death, that of Diversity most appropriately fits bacteria.

There is hardly a sterile place in this world. Even when people aim for a germ-free environment, the best we can manage is a room poor of germs. NASA assembles spacecraft in "clean rooms" they believed to be sterile, until microbiologists took a closer look. The results were shocking. All the major phyla were growing there: Proteobacteria, Firmicutes (including a number of spore formers), Actinobacteria, and Mollicutes. Some Euryarchaeota and Crenarchaeota were also detected. Three different clean rooms were shown to contain three different bacterial communities. Although these rooms were extremely dry, frequently treated with disinfectants, and very clean, so that little food was available for things to live there, they were anything but sterile. This investigation even led to a few newly discovered *Bacillus* species that have not yet been named. One can only wonder how many bacteria man has unknowingly sent into space, as inevitably spacecraft assembled in these "clean" rooms must have become contaminated.

Would bacteria survive a journey into space? Those unlucky enough to be on the outside of a spacecraft are unlikely to survive the extreme temperatures, vacuum, and radiation of cosmic rays and UV light. The atmosphere is the only earthly habitat that has not been mentioned yet. Without the help of humans, there are plenty of airborne bacteria present at low or high altitudes, blown there by the winds, and they can survive, although it is not unequivocally proven that they can also grow and divide. As they move further away from the earth, life becomes harder. The highest flyers were isolated at an altitude of 77 km. At a distance of 100 km from the earth's surface and beyond, the atmosphere is so thin that gases no longer protect against extreme temperatures, which can range from −100 to

1500°C, depending on the amount of solar radiation. Nevertheless, it would always feel very cold, by lack of heat-transferring gas molecules. The bombardment of cosmic rays increases as one moves further away from the earth, and these can be of high-energy or low-energy nature, but both are damaging to DNA and proteins. Organisms that can quickly repair such radiation, such as the *Deinococcus* species, would have an advantage here. It has been shown that *Deinococcus radiodurans* can easily repair the damage of cosmic radiation like that bombarding the surface of Mars, while the temperature of −80°C that would be typical at this planet actually reduces the radiation damage. (There is as yet no evidence that *Deinococcus*, or any other bacteria, for that sake, are actually living on Mars.) However, the combination of radiation, low temperatures and desiccation poses a major challenge for any living organism.

Inside a spacecraft, bacteria can reach outer space without much trouble. The high gravitational forces during takeoff and the lack of gravity once in space should not bother bacteria, according to some calculations, as their small size would make gravity a negligible force. Nevertheless, when this was experimentally tested, some bacteria grew slower inside a spaceship in orbit than in an earthly laboratory, while others seemed to grow normally. The difference was found to be due to motility. Lack of gravity is of no concern to motile bacteria with active flagella, as the cells can swim to reach their food, but nonmotile bacteria grow more slowly, presumably because they are surrounded by a motionless medium from which the nutrients have to be obtained by inefficient diffusion.

The survival capacity of bacteria exposed to space vacuum, with or without facing the sun, was tested experimentally for a range of days or months. Most cells died during these experiments; so far only a few organisms have been shown to survive a space trip. Surprisingly, some lichens (which are a symbiotic life form of algae and fungi) retained their photosynthetic ability. The halophilic (salt-living) marine Cyanobacterium *Synechococcus* could also survive a "space walk" of two weeks, provided it was put inside a salt crystal. Cyanobacteria seem to be capable of anything!

Such observations have fed speculations that bacteria could have been spread from one planet to another. Did life evolve on earth or did it arrive from some other planet, as the "panspermia" theory proposes? This question was not addressed in Chapter 3, where the origin of life was discussed. It would only transport the question of how life began to another planet, and would not solve the problem of producing living matter from nonliving precursor molecules. Nude spores are unlikely to survive an interplanetary journey, but deep inside asteroids or comets they might have a chance to survive, provided the trip did not take millions of years. It depends on the attitude of the reader to believe the likelihood of this scenario, which has not been proven or disproven. It is not known if our planet shares some of its microbes with other planets, irrespective of which was the planet of departure and which was the destination.

Maybe intelligent life studies bacteria on some other planet. Even if they do, their distance is hampering communication with us (provided that they wanted to communicate). More likely, though, their planet and ours are out of phase, so that intelligent life studied bacteria long before we had invented the microscope, or will do so long after we are gone. Humans will not always inhabit earth, as species come and go. But we can be certain that bacteria are here to stay, for as long as the earth will support life.

Glossary

AB_5 toxin A bacterial toxin consisting of six parts: one part of one type of protein and five parts of a second protein

Acidobacteria A phylum of the Eubacteria

Acidophiles Organisms living in acid conditions

Actinobacteria A phylum of the Eubacteria

Actinomycin An antibiotic discovered by S. Waksman, produced by *Streptomyces* bacteria. It inhibits protein synthesis

Adaptive immune system The part of the immune system that learns by experience, and produces antibodies. Vaccines work by triggering an adaptive immune response

AIDS Acquired Immune Deficiency Syndrome, the disease caused by HIV

Alkaliphiles Organisms living in alkali conditions (the opposite of acid conditions)

Amino acids The building blocks of proteins. There are 21 different amino acids

Anamox bacteria Recently discovered bacteria that can very efficiently convert ammonium to nitrogen gas

Anthrax The disease caused by *Bacillus anthracis*, also known as woolsorter's disease. It has a high mortality because of the toxins produced by the bacteria

Antibiotic A drug specifically targeting bacteria. Antibiotics either kill bacteria or inhibit their growth. They have no effect on viruses

Bacteria: The Benign, the Bad, and the Beautiful, First Edition. Trudy M. Wassenaar.

Antibiotic resistance See under "resistance"

Antibody A protein made by immune cells that binds specifically to bacteria or viruses. Antibodies are designed when the adaptive immune system encounters a pathogen for the first time, after which the cells remember how to produce these for future encounters

Archaea One of the three Domains of life, also called Archaebacteria. Archaea are prokaryotes and lack a nucleus, but their ribosomal RNA is different from that of eubacteria, and their membranes contain ether lipids

Atopic disease A disease caused by a malfunctioning immune system, such as allergies or eczema

ATP Short for adenosine triphosphate. It serves as the main energy carrier in the cell. It is also one of the building blocks of RNA

Autoimmune disease A disease caused by the immune system attacking the cells of the body

Autokinase An enzyme that can attach a phosphate group onto itself

Autotroph Bacteria that get their carbon from carbon dioxide

Bacteria One of the three Domains of life, also called Eubacteria. Bacteria are prokaryotes and lack a nucleus. Their membranes contain ester lipids, different from the lipids in membranes of Archaea and Eukaryotes

Bacteriochlorophyll A protein used by bacteria to harvest the energy of life, related to, but not identical to, chlorophyll of plants

Bacteriorhodopsin A bacterial light-harvesting protein structurally resembling rhodopsin in our eye

Bacteriophage A virus infecting bacteria or archaea

Bacteroidetes A phylum of the Eubacteria

Beta-lactamase Beta-lactam is a chemical structure that is part of the penicillin group of antibiotics, and beta-lactamase breaks down these antibiotics. Bacteria possessing a beta-lactamase gene are resistant to penicillins

Biofilm Growth of complex communities of sessile bacteria, which grow in layers covered with slime. Bacteria that form a biofilm communicate with each other

Bioinformatics A scientific discipline where biological information (mostly genetic information in the form of DNA or protein sequences) is researched with computers

Bioluminescence The property of living organisms to produce light with the help of enzymes, at the cost of energy

Bioremediation The removal of unwanted compounds, for example, chemical toxins or oil spills, by artificially applying bacteria that can degrade these

Biosafety levels Defined safety measures needed to work with particular pathogens or genetically modified organisms

Botox The neurotoxin produced by *Clostridium botulinum*, used in the cosmetic industry in very dilute concentrations to relax frown lines

Botulism The disease caused by *Clostridium botulinum*. It has a high mortality because of toxins produced by the bacteria. It can be present in home-canned food

Brownian motion The constant movement of cells under a microscope, the result of a bombardment of water molecules. First described by Robert Brown

BSE A cattle disease ("Mad cow disease") caused by a prion, an infectious, malfunctioning protein

Bt Short for *Bacillus thuringiensis* toxin, used as a pesticide

Chagas' disease The disease caused by unicellular eukaryotic *Trypanosoma cruzi*, spread by bites from infected *Triatoma* bugs

Carbohydrates One of the key components of living cells. Carbohydrates are made of sugar chains

Carbon fixation The process by which carbon dioxide from the air is converted into biomolecules, which requires a lot of energy. Often, this energy is provided via photosynthesis

Catalyst A chemical that speeds up a chemical reaction. Enzymes are biological catalysts

Cell wall The thick peptidoglycan polymer layer that surrounds the single membrane of Gram-positive bacteria. Plants have a cell wall, too, but that is composed of cellulose

Cellulase An enzyme that degrades cellulose, only produced by certain bacteria and fungi

Cellulose A strong carbohydrate polymer and constituent of the cell wall of plant cells. It is indigestible by most organisms, unless these contain cellulase, the enzyme that can digest cellulose

Channel protein Proteins in the membrane of a cell that let charged ions pass. Channel proteins can be in an open or closed position, which is regulated by the cell

Chaperone Proteins that bind to other proteins, thus helping them fold into the right configuration

Chemotroph Bacteria that get their energy from food in the form of high-energy molecules

Chlorobi A phylum of the Eubacteria

Chloroflexi A phylum of the Eubacteria

Chlorophyll A light-harvesting protein found in Cyanobacteria and in the chloroplasts of algae and plants

Chloroplasts The organelles in plant cells responsible for photosynthesis. Chloroplasts probably resulted from endosymbiotic Cyanobacteria

Cholera The disease caused by *Vibrio cholerae*. It has a high mortality because of cholera toxin and can cause large epidemics. It is transmitted via the oral–fecal route

Chromosome One complete, major DNA molecule of a cell. Many bacteria only have one chromosome, though some species have two or three different chromosomes. Eukaryotes usually have multiple chromosomes

Clone (i) The offspring of a single (bacterial) cell, resulting in a monoculture of identical cells. (ii) A genetic copy of a eukaryotic individual

Commensals Bacteria that live in or on a host without doing harm, or even being beneficial

Crenarchaeota A phylum of the Archaea

Cyclic AMP A molecule in the cell related to ATP. It is a general messenger and is essential in the regulation of electrolyte concentrations; the membrane channels are regulated via cyclic AMP

Defensins Short peptides produced by macrophages to kill bacteria

Deinococcus-Thermus A phylum of the Eubacteria

Denitrifying bacteria Bacteria that convert nitrogen present in biomass or nitrate into nitrogen gas

Differentiation Cells undergo differentiation when they vary in shapes and function depending on their location in a multicellular organism. Differentiation of bacterial cells is only observed Myxobacteria and some Cyanobacteria

Diphtheria The disease caused by *Corynebacterium diphtheriae*. When common, it caused high mortality in children because of a toxin. Nowadays, the disease is uncommon due to effective vaccination

Dispersant A chemical that disperses oil into small droplets that can float in water

DNA fingerprint A banding pattern that is generated from DNA using PCR in such a manner that the pattern is unique to an individual

DNA polymerase The enzyme that produces DNA from the nucleotide building blocks (G, A, T, and C) in the cell

DNAse An enzyme that degrades DNA

Domain The highest unit of the tree of life; currently three Domains are recognized: the Archaea, Bacteria (also called Eubacteria), and Eukaryotes

Elephantiasis A disease caused by parasitic worms, spread by infected mosquitoes

Endosymbionts Bacteria that live inside other cells, either as parasites or as commensals

Enzyme A protein with a catalytic function. Most proteins are enzymes, and all processes in the cell are performed with the help of enzymes

Epidemic A large outbreak of an infectious disease

Erythrocytes Red blood cells that contain hemoglobin, the protein that transports oxygen

Erythromycin An antibiotic belonging to the macrolide group

Eubacteria One of the three Domains of life, also called Bacteria. They are prokaryotes and lack a nucleus, but their ribosomal RNA is different from that of archaea, and their membranes contain ester lipids

Eukaryotes One of the three Domains of life. Their cells contain a nucleus and specialized organelles such as mitochondria. Eukaryotes can be unicellular or multicellular organisms

Euryarchaeota A phylum of the Archaea

Exon A segment of a eukaryotic gene on which protein-coding information is stored. Many eukaryotic genes consist of several exons, separated by noncoding DNA segments called introns

Extremophiles Bacteria or archaea that live under extreme conditions

Firmicutes A phylum of the Eubacteria

Flagella Appendices on a bacterial or archaeal body that produce motility through rotation, consisting of a long tail and a motor complex

Flagellin The protein that builds the flagellar tail

Fusobacteria A phylum of the Eubacteria

GC-rich DNA DNA that contains more of the bases G and C than it contains the bases A and T. Bacteria can be taxonomically divided according to their G-C content, but G-C content is not strictly conserved within bacterial phyla

Gene A segment of DNA that bears the information to produce a particular product, which is either an RNA molecule, or a protein (produced by a messenger RNA intermediate). The information is stored in the order, or sequence, of the four building blocks of DNA

Genetic code The code by which a DNA sequence of nucleotides is translated into a sequence of amino acids to form a protein. Each triplet (three nucleotide bases) code for one particular amino acid

Genetics A scientific discipline that studies genes and their products

Genome All DNA found in a cell is called the genome. This can consist of a single chromosome, multiple chromosomes or chromosome plus plasmid(s)

Genomics A scientific discipline that investigates and compares the genome sequences of organisms

Genus The lowest taxonomic group of related species, given by the first name of a full biological species name

Gliding Motility by means of slime secretion, practiced by Cyanobacteria

Glycosylation The process of attaching a sugar group to another molecule, often a protein

Gram-negative Bacteria that color violet in a Gram stain are called Gram-negative. They have two membranes

Gram-positive Bacteria that color purple in a Gram stain are called Gram-positive. They have a single membrane surrounded by a peptidoglycan cell wall

Gram stain A staining procedure designed by Hans Christian Gram

Green fluorescent protein, GFP A bioluminescent protein isolated from jellyfish, frequently used as a biomarker

Hemoglobin The protein in red blood cells that binds oxygen and gives the cells their color

Hemolysin A toxin that can lyse erythrocytes (red blood cells)

Hemolysis A laboratory test to demonstrate the activity of a pore-forming toxin, by its ability to lyse erythrocytes

Haloalkaliphiles Bacteria or archaea that live in salty, alkaline conditions

Halophiles Bacteria or archaea that live in high salt conditions

Heartwater disease A cattle disease caused by *Ehrlichia ruminantium* that is transmitted through ticks

Heat shock protein A chaperone protein that binds to other proteins that were damaged by stress, such as heat, in order to refold these or else determine them for degradation. Heat shock proteins are overexpressed during stress; hence, they are also called stress proteins

Heterotroph Bacteria that get their carbon from complex molecules (as opposed to carbon dioxide from the air)

Homeostasis The processes taking place in the cell to remain healthy, and repair damage

Host An organism on or in which bacteria live, either as commensals or as pathogens

Host jump A virus or bacterium that used to be adapted to one host can change to another host, frequently causing disease in the second

Host–microbe interactions The molecular processes taking place in the gut where host cells and bacteria sense each other and respond accordingly

Hydrocarbons Organic compounds composed of hydrogen and carbon, with some minor elements. Oil is a mixture of hydrocarbons

Inflammatory bowel disease, IBD A chronic gut disease, believed to be caused by incorrect interaction of gut cells with the microflora

Immune system A complex interplay of body cells to combat viral and bacterial infections. The immune system will attack any cells that are recognized as "foreign" as these can be distinguished from own body cells

Innate immune system The part of the immune system that is continuously active, without the need for prior exposure to a pathogen

Insulin A hormone produced by the pancreas that regulates blood sugar levels

Interleukin Messenger compounds produced by immune cells to send signals to other immune cells

Intron A section of a eukaryotic gene that interrupts the protein-coding exons. Many eukaryotic genes are interrupted by multiple introns, but few bacterial genes contain introns

Kinase An enzyme that links a phosphate group to a substrate

Kingdom A division in taxonomy

Korarchaeota A phylum of the Archaea

Lactose A sugar found in milk

Lactase The enzyme needed to degrade lactose. Lactase is normally expressed in young mammals only. In humans, a proportion of the population carries a mutation that produces lactase throughout life, so that milk is tolerated even by adults. Lactose intolerance is common in Asia

Leishmaniasis The disease caused by unicellular eukaryotic *Leishmania* species, spread by infected sandflies

Leptospirosis The disease caused by *Leptospira* species, transmitted by the urine of infected rodents. Also known as Weil's disease

Lipid A biomolecule that consists of long carbohydrate chains, of which membranes are composed

Lipase An enzyme that breaks down lipids

LPS Abbreviation of Lipopolysaccharide, a constituent of the outer membrane of Gram-negative bacteria. It consists of a lipid part coupled to a chain of sugar parts

Luciferase An enzyme that emits light because of an energy-consuming oxidation reaction

Lyme disease The disease caused by *Borrelia burgdorferi*, also called borreliosis. It is transmitted by ticks

Lysis A cell is lysed when its membrane disintegrates and the cell content leaks out

Lysozyme An enzyme degrading bacterial cell walls, resulting in lysis of the targeted cells

Macrolide A group of antibiotics that inhibit protein synthesis

Macrophage Specialized immune cell dedicated to engulf bacteria and kill these

Malaria The disease caused by unicellular eukaryotic *Plasmodium* species, spread by bites from infected *Anopheles* mosquitoes

Metabolism The combined processes of degrading and building complex compounds in the cell during life and growth

Metagenomics A scientific discipline that characterizes all DNA in a given biotope

Metalloprotease An enzyme that degrades proteins and needs one or more metal atoms in order to function

Methanogens Bacteria that produce methane

Methicillin An antibiotic of the penicillin group

Microbiome A combination of all bacterial genomes that live in a particular ecological niche, for instance, the microbiome of the human gut

Microflora The collective term for all bacteria living in a particular niche, for instance, the oral microflora

Milky disease A disease of the larvae of Japanese Beetle caused by *Bacillus popillae*

Mitochondria The organelles in eukaryotic cells specialized to provide the cell with energy. Mitochondria probably resulted from endosymbiotic Proteobacteria

Mobile element A piece of DNA that can move itself from one DNA location to another

Mollicutes A phylum of the Eubacteria

Morphology The study of (body) shape

MRSA Methicillin resistant *Staphylococcus aureus*, a pathogen that frequently causes hospital infections

Mutation, mutant A mutation is a change in a DNA sequence, which may or may not lead to a phenotype (a noticeable change in a characteristic). The bacteria in which a mutation has occurred is called a mutant. A mutant gene or a mutant strain is derived from the naturally found, or wild-type, variant

Mutualism A life form where two organisms live together, to each other's benefit, which can also live independently

Myxospore A specialized cell form of Myxobacteria, which forms when conditions do not favor growth

Nanoarchaeota A phylum of the Archaea

Nanobacteria A group of very small bacteria (not a taxonomic term)

Nucleus The partitioning in a eukaryotic cell where most of the DNA is stored, surrounded by a membrane

Nucleotide Building block of DNA or RNA. There are four different nucleotides; in DNA, these are abbreviated using A, G, C and T. In RNA, the T is replaced by U. Each nucleotide contains a base, and "base" is used to express DNA length (as in kb)

Nitrogenase The crucial enzyme involved in nitrogen fixation

Nitrogen fixation The process by which nitrogen gas from the air is converted into biomolecules, a process that requires a lot of energy and can only take place in the absence of oxygen

Opportunistic pathogen A pathogen that can only cause disease when specific conditions apply in the host, usually a weakened immune system

Pandemic An epidemic involving several countries or continents

Panspermia The theory describing that life did not originate on Earth but was seeded from some other planet via asteroids or comets

Parallel evolution The process by which a product results twice via independent evolutionary events

Parasite (i) An organism that lives in a host, at the expense of that host. (ii) Eukaryotic pathogens such as worms or protozoa

Pathogen A bacterium, virus, or eukaryote that causes disease

Pathogenesis The manner in which a pathogen causes disease

Pattern recognition receptor, PRR Receptors on the outside of immune cells that recognize conserved patterns rather than unique structures, as part of the innate immune system

PCR Polymeric Chain Reaction, a technique to produce multiple copies of an exactly defined piece of DNA, using *Taq* polymerase

Penicillin A group of antibiotics of the beta-lactams, that act by inhibiting the formation of a cell wall. Penicillin was first discovered by A. Fleming

Peptide A very short chain of amino acids

Peptidoglycan The constituent of the cell wall of Gram-positive bacteria. Gram-negative bacteria only have small amounts of peptidoglycan and do not have a cell wall

Phage Short for bacteriophage, a virus infecting bacteria

Phenotype An observable characteristic, resulting from a particular gene or set of genes. A phenotype can change when a gene is added or removed, or when it is activated or inactivated as a result of mutation

Phospholipase An enzyme breaking down phospholipids, constituents of membranes

Phospholipids Constituents of membranes

Photosynthesis The process by which bacteria and plants can use energy from light and convert this into chemical energy. The energy is then used to produce sugar from carbon dioxide

Phototroph Bacteria that get their energy from light

Phylum A main division in taxonomy, the highest branch in the bacterial world

Pili Appendices on a bacterial or archaeal body that bacteria use to adhere to a surface, sometimes used to move along that surface

Plague The disease caused by *Yersinia pestis*. It has a high mortality because of invading bacteria that cause septic shock. It is transmitted from infected rats via fleas

Planctomycetes A phylum of the Eubacteria

Plasmid A DNA molecule that exists independent of the chromosome in the bacterial cell, and replicates independently. A plasmid is often smaller than a chromosome and may carry antibiotic resistance genes

Plasmodium A protozoa causing malaria. It is transmitted by Anopheles mosquitoes

Poliomyelitis virus The virus that causes polio. Attempts to eradicate the disease globally by vaccination programs are ongoing

Pore A hole in a membrane, typically produced by toxins, which will cause the cell to die. (Sometimes the term is used to describe the natural channels in a cell's membrane, and the proteins forming these are called porins)

Pore-forming toxin A toxin that works by making holes in the target membrane

Prion A wrongly-folded protein that is infectious, in that it forces correct protein molecules to refold into a prion as well. BSE is caused by prions

Probiotics Intestinal bacteria that are beneficial to health

Protease An enzyme that breaks down proteins

Protein One of the key components of living cells. Proteins are made of amino acids chains, which fold into a three-dimensional configuration to be active in the cell

Protein kinase An enzyme that attaches a phosphate group to a protein

Protein phosphorylation The process of phosphate groups being attached to proteins by the action of specific kinases

Protein synthesis The process by which the cell produces proteins from messenger RNA, with the use of ribosomes (amongst other components). Amino acids are combined in the correct order by the ribosome, dictated by the genetic code on the RNA. Antibiotics and bacterial toxins frequently interfere with protein synthesis

Proteobacteria A phylum of the Eubacteria

Protozoa The collective group of single-cell eukaryotic organisms, excluding fungi

Psychrophiles Organisms living in cold conditions

Quorum sensing The process by which bacteria living in a biofilm sense cell density, and react to this density

Recombinant DNA DNA that contains genes in a combination that does not occur naturally, but is produced in the laboratory

Replication The production of a DNA copy, required before a cell can divide

Reporter gene A gene that can report whether it is used by the cell, for instance, by producing a pigment

Resistance Short for acquired antibiotic resistance. It describes bacteria that can grow in the presence of an antibiotic to which they would normally be susceptible. Resistance can result from gene uptake, DNA mutations, or changes in gene expression

Ribosomal RNA A particular kind of RNA that builds the ribosome, together with a number of proteins. Its gene is used in taxonomy to determine the relationship between species

Ribosome The functional entity of a cell responsible for protein synthesis. The ribosome consists of ribosomal RNA and proteins; The ribosome slides along a messenger RNA to read the genetic code with the help of transfer-RNA molecules, and attaches the correct amino acids to a growing protein chain

River blindness The disease caused by the roundworm *Onchocerca volvulus*, spread by bites from infected blackflies

Rubisco The key enzyme in carbon fixation. Ribulose-1,5-bisphosphate carboxylase/oxygenase is found in plants, algae, Cyanobacteria, and other autotrophic bacteria. It binds CO_2 to ribulose-1,5-bisphosphate, producing two 3-phosphoglycerate molecules, one of which will be incorporated into sugar

Shiga toxin The toxin produced by *Shigella dysenteriae*. It inhibits protein synthesis by interfering with the ribosomes and is composed of five subunits of one protein, and one subunit of another, with catalytic activity (AB_5 structure)

Shigellosis The disease caused by *Shigella* species. It has a low mortality and is a food-borne disease

Sleeping sickness The disease caused by unicellular eukaryotic *Trypanosoma brucei*, spread by bites from infected tsetse flies

Sliding Bacterial motility by means of pili, more specifically Type IV pili

Species The lowest taxonomic unit, given by the second name of a full biological species name. In animals and plants, individuals of a species can breed and produce fertile offspring. Bacterial species are defined by the characteristics of the cells

Spirochetes A phylum of the Eubacteria

Spore Short for endospore. A dormant form of bacteria, often Firmicutes, that can survive hostile environments for a long time. A spore is not alive as it does not repair damage, but when it encounters favorable conditions it can change back to a growing cell

Stress response The combined changes of gene expression observed in stressed cells, for instance, after a heat or cold shock

Streptomycin An antibiotic discovered by S. Waksman, produced by *Streptomyces* bacteria and belonging to the group of aminoglycosides. It inhibits protein synthesis

Stromatolites Marine, mushroom-shaped formations of Cyanobacteria that grow in coastal areas

Superbug Informal term for bacteria with multiple antibiotic resistances that cause infections that are hard to treat

Syphilis The disease caused by *Treponema pallidum*, a sexually transmitted disease

***Taq* polymerase** *Taq* is the abbreviation of *Thermus aquaticus*, a Gram-negative thermophile with an optimal growth temperature of 70°C. Its DNA polymerase is used for PCR

Taxonomy A scientific discipline to order organisms and establish evolutionary relationships

Tetanus The disease caused by *Clostridium tetani*, also known as lockjaw. It has a high mortality because of an extremely potent toxin. It is caused by bacteria entering a deep wound

Thermophiles Bacteria that grow at high temperatures

Thermoacidophiles Bacteria that grow at high temperatures and in acid conditions

Transposon A kind of mobile, parasitic DNA that can insert itself in the genome of bacteria. It frequently carries antibiotic resistance genes

Trypanosoma A protozoa causing sleeping sickness, a tropical disease. It is transmitted by tsetse flies

Tuberculosis The disease caused by *Mycobacterium tuberculosis*. It has a low mortality (higher in HIV-infected patients) and is spread by aerosols

Two-component signal transduction system A system by which bacteria sense changes in their environment, which they transfer to a signal inside the cell to respond accordingly

Type IV pili A type of pili, hair-like appendages on a bacterial body, produced by some bacteria. These can be involved in motility (sliding) or in uptake of DNA. In some cases, the pili are used for secretion of proteins

Type Three Secretion System, TTSS Appendices on a bacterial or archaeal body that are used as an injection needle, to inject effector proteins into a eukaryotic target cell. They are frequently part of a pathogenic strategy

Typhus The disease caused by *Rickettsia prowazekii*. It has a medium mortality rate and is transmitted via infected lice, either through their bites or via their feces. Not to be confused with typhoid fever, a food-borne disease that is caused by *Salmonella typhi*

Unicellular The lifestyle of prokaryotes or eukaryotes where single cells live independently

Upregulation A gene is upregulated when more messenger RNA, and thus more of its protein product, is formed in a cell

Vector (i) An animal host by which an infectious disease is spread, usually an insect or a tick. (ii) An independently replicating piece of DNA used for cloning, for example, a plasmid

Virulence gene A gene whose product is involved in pathogenicity: the property to cause disease

Virus A nonliving biological entity consisting of DNA or RNA (never both) and protein, sometimes covered by a membrane. A virus can only multiply intracellularly. Viruses often cause disease but can also remain in a cell unnoticed. Many viruses incorporate their DNA into the host cell, and can get trapped in there when they have lost the ability to excise, as a result of mutations. This is called a provirus. Viruses parasitizing on bacteria are called bacteriophages or phages

Wild-type Of the type that occurs in Nature, as opposed to mutant variants introduced by DNA changes. The term can apply to a gene or to a bacterial strain

Whooping cough The disease caused by *Bordetella pertussis*. It has a moderate mortality and is spread by aerosols

Zoonosis An infectious disease that humans get from animals

Bibliography

Abstracts from cited articles are available from: http://www.ncbi.nlm.nih.gov

CHAPTERS 1 AND 2

Aziz RK. The case for biocentric microbiology. Gut 2009; 1(1):16.

Cavalier-Smith T. Cell evolution and earth history: stasis and revolution. Philos Trans R Soc Lond B Biol Sci 2006; 361(1670):969–1006.

Davidov Y, Jurkevitch E. Predation between prokaryotes and the origin of eukaryotes. Bioessays 2009; 31(7):748–757.

El Albani A, et al. Large colonial organisms with coordinated growth in oxgenated environments 2.1 Gyr ago. Nature; 466(7302):100–104.

History and scope of Medical Microbiology. part 1: History of Medical Microbiology. Jaypee Brothers. Available at http://www.jaypeebrothers.com/pdf/his_med_bio.pdf. Last accessed date: 9 September 2011

Whitman WB, Coleman DC, Wiebe WJ. Prokaryotes: the unseen majority. Proc Natl Acad Sci U S A 1998; 95(12):6578–6583.

CHAPTER 3

Amel BK, Amine B, Amina B. Survival of *Vibrio fluvialis* in seawater under starvation conditions. Microbiol Res 2008; 163(3):323–328.

Foster GT, editor. Focus on Bioterrorism. New York: Nova Science Publishers, Inc.; 2006.

Johnson SS, et al. Ancient bacteria show evidence of DNA repair. Proc Natl Acad Sci U S A 2007; 104(36):14401–14405. Erratum in: Proc Natl Acad Sci U S A. 2007; 104(51):20635 and in Proc Natl Acad Sci U S A. 2008; 105(30):10631.

CHAPTER 4

Adams DG. How do *Cyanobacteria* glide? Microbiology Today 2001; 28:131–133.

Jarrell KF, McBride MJ. The surprisingly diverse ways that prokaryotes move. Nat Rev Microbiol 2008; 6(6):466–476.

Terashima H, Kojima S, Homma M. Flagellar motility in bacteria structure and function of flagellar motor. Int Rev Cell Mol Biol 2008; 270:39–85.

Trachtenberg S, Cohen-Krausz S. The archaeabacterial flagellar filament: a bacterial propeller with a pilus-like structure. J Mol Microbiol Biotechnol 2006; 11(3–5):208–220.

CHAPTER 5

Enninga J, Rosenshine I. Imaging the assembly, structure and activity of type III secretion systems. Cell Microbiol 2009; 11(10):1462–1470.

Moraes TF, Spreter T, Strynadka NC. Piecing together the type III injectisome of bacterial pathogens. Curr Opin Struct Biol 2008; 18(2):258–266.

CHAPTER 7

Nobel prize web site: http://nobelprize.org.

Gross L. How Charles Nicolle of the Pasteur Institute discovered that epidemic typhus is transmitted by lice. Proc Natl Acad Sci U S A 1996; 93(20):10539–10540.

Mariani SM. Spongiform encephalopathies: a tale of cannibals, cattle, and prions. Med Gen Med 2003; 5(3):42.

WHO report: Global burden of disease 2004. Part 2: Causes of death http://www.who.int/healthinfo/global_burden_disease/GBD_report_2004update_part2.pdf.

CHAPTER 8

Aksoy S, Rio RV. Interactions among multiple genomes: tsetse, its symbionts and trypanosomes. Insect Biochem Mol Biol 2005; 35(7):691–698.

Dess A, et al. *Serratia marcescens* infections and outbreaks in neonatal intensive care units. J Chemother 2009; 21(5):493–499.

Kudo T. Termite-microbe symbiotic system and its efficient degradation of lignocellulose. Biosci Biotechnol Biochem 2009; 73(12):2561–2567.

Lehane MJ. The biology of blood-sucking insects. Cambridge UK: Cambridge University Press, 2005.

Martin C, Gavotte L. The bacteria *Wolbachia* in filariae, a biological Russian dolls' system: new trends in antifilarial treatments. Parasite 2010; 17(2):79–89.

Ramsey JS, et al. Genomic evidence for complementary purine metabolism in the pea aphid, *Acyrthosiphon pisum*, and its symbiotic bacterium *Buchnera aphidicola*. Insect Mol Biol 2010; 19(Suppl 2):241–248.

Russell JE, Stouthamer R. The genetics and evolution of obligate reproductive parasitism in *Trichogramma pretiosum* infected with parthenogenesis-inducing *Wolbachia*. Heredity 2010; 106(1):58–67.

Sharon G, et al. Commensal bacteria play a role in mating preference of *Drosophila melanogaster*. Proc Natl Acad Sci U S A 2010; 107(46):20051–20056.

Skotarczak B. Canine ehrlichiosis. Ann Agric Environ Med 2003; 10(2):137–141.

Tilly K, Rosa PA, Stewart PE. Biology of infection with *Borrelia burgdorferi*. Infect Dis Clin North Am 2008; 22(2):217–234.

Tolle MA. Mosquito-borne diseases. Curr Probl Pediatr Adolesc Health Care 2009; 39(4):97–140.

CHAPTER 9

Cowden JM, Ahmed S, Donaghy M, Riley A. Epidemiological investigation of the central Scotland outbreak of *Escherichia coli* O157 infection, November to December 1996. Epidemiol Infect 2001; 126(3):335–341.

Koussoulakos S. Botulinum neurotoxin: the ugly duckling. Eur Neurol 2009; 61(6): 331–342.

Moayeri M, Leppla SH. Cellular and systemic effects of anthrax lethal toxin and edema toxin. Mol Aspects Med 2009; 30(6):439–455.

Pennington H. When food kills. BSE, *E. coli* and disaster science. New York: Oxford University Press Inc; 2003.

Popoff MR, Bouvet P. Clostridial toxins. Future Microbiol 2009; 4(8):1021–1064.

Salyers AA, Whitt DD. Bacterial pathogenisis. A molecular approach, Washington DC: ASM Press; 1994.

Sandvig K, et al. Protein toxins from plants and bacteria: probes for intracellular transport and tools in medicine. FEBS Lett 2010; 584(12):2626–2634.

Tang WJ, Guo Q. The adenylyl cyclase activity of anthrax edema factor. Mol Aspects Med 2009; 30(6):423–430.

Tuttle J, et al. Lessons from a large outbreak of *Escherichia coli* O157:H7 infections: insights into the infectious dose and method of widespread contamination of hamburger patties. Epidemiol Infect 1999; 122(2):185–192.

CHAPTER 10

Aehle W, editor. Enzymes in industry. Production and application. 3rd ed. Weinheim: Wiley-VCH; 2007.

Bhosale SH, Rao MB, Deshpande VV. Molecular and industrial aspects of glucose isomerase. Microbiol Rev 1996; 60(2):280–300.

Rabinow P. Making PCR: a story of biotechnology. Chicago: The University of Chicago Press; 1996.

CHAPTER 11

Falkow S. Molecular Koch's postulates applied to microbial pathogenicity. Rev Infect Dis 1988; 10(Suppl 2):S274–S276.

Falkow S. Molecular Koch's postulates applied to bacterial pathogenicity–a personal recollection 15 years later. Nat Rev Microbiol 2004; 2(1):67–72.

Ghim CM, Lee SK, Takayama S, Mitchell RJ. The art of reporter proteins in science: past, present and future applications. BMB Rep 2010; 43(7):451–460.

Lagesen K, Ussery DW, Wassenaar TM. Genome update: the 1000th genome—a cautionary tale. Microbiology 2010; 156(3):603–608.

Ussery DW, Borini S, Wassenaar TM. Bioinformatics for microbiologists. Computing for Comparative Microbial Genomics series. UK: Springer; 2008.

CHAPTER 12

Hoban DJ. Antibiotics and collateral damage. Clin Cornerstone 2003; 3(Suppl):S12–S20.

Kumarasamy KK, et al. Emergence of a new antibiotic resistance mechanism in India, Pakistan, and the UK: a molecular, biological, and epidemiological study. Lancet Infect Dis 2010; 10(9):597–602.

Lan R, Reeves PR, Octavia S. Population structure, origins and evolution of major *Salmonella enterica* clones. Infect Genet Evol 2009; 9(5):996–1005.

Velayati AA, et al. Emergence of new forms of totally drug-resistant tuberculosis bacilli: super extensively drug-resistant tuberculosis or totally drug-resistant strains in Iran. Chest 2009; 136(2):420–425.

Winstanley TG, Spencer RC. Antibiotic dependence in a strain of *Neisseria pharyngis*. J Hosp Infect 1987; 10(1):87–90.

CHAPTER 13

Bryant DA, Frigaard NU. Prokaryotic photosynthesis and phototrophy illuminated. Trends Microbiol 2006; 14(11):488–496.

Eiler A. Evidence for the ubiquity of mixotrophic bacteria in the upper ocean: implications and consequences. Appl Environ Microbiol 2006; 72(12):7431–7437.

Herndl GJ, et al. Contribution of Archaea to total prokaryotic production in the deep Atlantic Ocean. Appl Environ Microbiol 2005; 71(5):2303–2309.

Munn CB. Marine Microbiology. Ecology and applications. Oxon UK: Garland Science/BIOS Scientific Publishers; 2004.

Woese CR, Fox GE. Phylogenetic structure of the prokaryotic domain: the primary kingdoms. Proc Natl Acad Sci U S A 1977; 74(11):5088–5090.

Venter JC, et al. Environmental genome shotgun sequencing of the Sargasso Sea. Science 2004; 304(5667):66–74.

CHAPTER 14

Brown LR. Microbial enhanced oil recovery (MEOR). Curr Opin Microbiol 2010; 13(3):316–320.

Coleman J, Chair. Oil in the Sea III: Inputs, Fates and Effects National Research Council of the National Academies, Washington, DC: The National Academies Press; 2003. Available at www.nap.edu.

Jones DM, et al. Crude-oil biodegradation via methanogenesis in subsurface petroleum reservoirs. Nature 2008; 451(7175):176–180.

Kumar A, et al. Enhanced CO_2 fixation and biofuel production via microalgae: recent developments and future directions. Trends Biotechnol 2010; 28(7):371–380.

Tabita FR, et al. Distinct form I, II, III, and IV Rubisco proteins from the three kingdoms of life provide clues about Rubisco evolution and structure/function relationships. J Exp Bot 2008; 59(7):1515–1524.

Vila J, et al. Microbial community structure of a heavy fuel oil-degrading marine consortium: linking microbial dynamics with polycyclic aromatic hydrocarbon utilization. FEMS Microbiol Ecol 2010; 73(2):349–362.

Yakimov MM, Timmis KN, Golyshin PN. Obligate oil-degrading marine bacteria. Curr Opin Biotechnol 2007; 18(3):257–266.

CHAPTER 15

Aiello AE, Larson EL, Levy SB. Consumer antibacterial soaps: effective or just risky? Clin Infect Dis 2007; 45(Suppl 2):S137–S147.

Chivian D, et al. Environmental genomics reveals a single-species ecosystem deep within earth. Science 2008; 322(5899):275–278.

Gilbert JA, Hill PJ, Dodd CER, Laybourn-Parry J. Demonstration of antifreeze protein activity in Antarctic lake bacteria. Microbiol 2004; 150(pt 1):171–180.

Horikoshi K. Alkaliphiles: some applications of their products for biotechnology. MMBR 1999; 63(4):735–750.

Huber H, Hohn MJ, Stetter KO, Rachel R. The phylum Nanoarchaeota: present knowledge and future perspectives of a unique form of life. Res Microbiol 2003; 154(3):165–171.

Löhr AJ, et al. Microbial communities in the world's largest acidic volcanic lake, Kawah Ijen in Indonesia, and in the Banyupahit river originating from it. Microb Ecol 2006; 52(4):609–618.

Mason OU, et al. First investigation of the microbiology of the deepest layer of ocean crust. PLoS One 2010; 5(11):e15399.

CHAPTER 16

Ben-Jacob E. Social behavior of bacteria: from physics to complex organization. Eur Phys J B 2008; 65(3):315–322.

Claverie JM, Abergel C. Mimivirus: the emerging paradox of quasi-autonomous viruses. Trends Genet 2010; 26(10):431–437.

Den Boer JW, et al. A large outbreak of Legionnaires' disease at a flower show, the Netherlands, 1999. Emerg Infect Dis 2002; 8(1):37–43.

Ellin Doyle M. Survival and growth of *Clostridium perfringens* during the cooling step of thermal processing of meat products. Madison, WI: Food Research Institute, University of Wisconsin; 2002 Feb. Available at www.amif.org/ht/a/GetDocumentAction/i/7434. Last accessed date: 9 September 2011.

Fischer MG, Allen MJ, Wilson WH, Suttle CA. Giant virus with a remarkable complement of genes infects marine zooplankton. Proc Natl Acad Sci U S A 2010; 107(45):19508–19513.

Nadell CD, Xavier JB, Foster KR. The sociobiology of biofilms. FEMS Microbiol Rev 2009; 33(1):206–224.

Nadell CD, Xavier JB, Levin SA, Foster KR. The evolution of quorum sensing in bacterial biofilms. PLoS Biol 2008; 6(1):e14.

Schulz HN, Jorgensen BB. Big bacteria. Size and volume of the largest prokaryotes. Annu Rev Microbiol 2001; 55:105–137.

Sockett RE. Predatory lifestyle of *Bdellovibrio bacteriovorus*. Annu Rev Microbiol 2009; 63:523–539.

Vijayachari P, Sugunan AP, Shriram AN. Leptospirosis: an emerging global public health problem. J Biosci 2008; 33(4):557–569.

CHAPTER 17

Microbial art website: http://www.microbialart.com.

http://www.notablebiographies.com.

http://www.smithsonianmag.com/science-nature/Painting-With-Penicillin-Alexander-Flemings-Germ-Art.html. Last accessed date: 9 September 2011

http://star.tau.ac.il/~eshel/image-flow.html.

Bastian F, et al. The microbiology of Lascaux Cave. Microbiology 2010; 156(Pt 3):644–652.

Bäzner H, Hennerici MG. Syphilis in German-speaking composers—examination results are confidential. Front Neurol Neurosci 2010; 27:61–83.

Ben-Jacob E. From snowflake formation to growth of bacterial colonies II: Cooperative formation of complex colonial patterns. Contemp Phys 1997; 38(3):205–241.

Cappitelli F, Principi P, Sorlini C. Biodeterioration of modern materials in contemporary collections: can biotechnology help? Trends Biotechnol 2006; 24(8):350–354.

Cappitelli F, Sorlini C. From papyrus to compact disc: the microbial deterioration of documentary heritage. Crit Rev Microbiol 2005; 31(1):1–10.

Chalke HD. The impact of tuberculsosis on history, literature and art. Med Hist 1962; 6(4):301–318.

Charkoudian LK, Fitzgerald JT, Khosla C, Champlin A. In living color: bacterial pigments as an untapped resource in the classroom and beyond. PLoS Biol 2010; 8(10):e1000510.

Franzen C. Syphilis in composers and musicians–Mozart, Beethoven, Paganini, Schubert, Schumann, Smetana. Eur J Clin Microbiol Infect Dis 2008; 27(12):1151–1157.

Pepe O, et al. Heterotrophic microorganisms in deteriorated medieval wall paintings in southern Italian churches. Microbiol Res 2010; 165(1):21–32.

Santos A, et al. Application of molecular techniques to the elucidation of the microbial community structure of antique paintings. Microb Ecol 2009; 58(4):692–702.

Tcherpakov M, Ben-Jacob E, Gutnick DL. *Paenibacillus dendritiformis sp. nov.*, proposal for a new pattern-forming species and its localization within a phylogenetic cluster. Int J Syst Bacteriol 1999; 49(Pt 1):239–246.

CHAPTER 18

Deutsch C, et al. Spatial coupling of nitrogen inputs and losses in the ocean. Nature 2007; 445(7124):163–167.

Kumar K, Mella-Herrera RA, Golden JW. Cyanobacterial heterocysts. Cold Spring Harb Perspect Biol 2010; 2(4):a000315.

Leigh JA. Nitrogen fixation in Methanogens: the archaeal perspective. Curr Issues Mol Biol 2000; 2(4):125–131.

Payne WJ. Centenary of the isolation of denitrifying bacteria. ASM News 1986; 52(12):627–629.

Setubal JC, et al. Genome sequence of *Azotobacter vinelandii*, an obligate aerobe specialized to support diverse anaerobic metabolic processes. J Bacteriol 2009; 191(14):4534–4545.

Tavares P, Pereira AS, Moura JJ, Moura I. Metalloenzymes of the denitrification pathway. J Inorg Biochem 2006; 100(12):2087–2100.

Wichter C, et al. Archaeal nitrification in the ocean. Proc Natl Acad Sci U S A 2006; 103(33):12317–12322.

CHAPTER 19

Amerithrax investigative summary. The United States Department of Justice. 2010 Feb 19. Available at www.justice.gov/amerithrax/docs/amx-investigative-summary.pdf.

Ayyadurai S, Sebbane F, Raoult D, Drancourt M. Body lice, *Yersinia pestis orientalis*, and black death. Emerg Infect Dis 2010; 16(5):892–893.

Chaignat C. What about cholera vaccines? Expert Rev Vaccines 2008; 7(4):403–405.

Faruque SM, Albert MJ, Mekalanos JJ. Epidemiology, genetics, and ecology of toxigenic *Vibrio cholerae*. Microbiol Mol Biol Rev 1998; 62(4):1301–1314.

Koch T, Denike K. Crediting his critics' concerns: remaking John Snow's map of Broad Street cholera, 1854. Soc Sci Med 2009; 69(8):1246–1251.

Littman RJ. The plague of Athens: epidemiology and paleopathology. Mt Sinai J Med 2009; 76(5):456–467.

Morens DM. Epidemic anthrax in the eighteenth century, the Americas. Emerg Infect Dis 2002; 8(10):1160–1162.

Palmer PES, Reeder MM, Editors. The imaging of tropical diseases: with epidemiological, pathological and clinical correlation. Volume 21. Berlin: Springer; 1981.

Raoult D, et al. Molecular identification by suicide PCR of *Yersinia pestis* as the agent of medieval black death. Proc Natl Acad Sci U S A 2000; 97(23):12800–12803.

Senderovich Y, Izhaki I, Halpern M. Fish as reservoirs and vectors of *Vibrio cholerae*. PLoS One 2010; 5(1):e8607.

Zhou D, Yang R. Molecular Darwinian evolution of virulence in *Yersinia pestis*. Infect Immun 2009; 77(6):2242–2250.

CHAPTER 20

Cani PD, Delzenne NM. Interplay between obesity and associated metabolic disorders: new insights into the gut microbiota. Curr Opin Pharmacol 2009; 9(6):737–743.

Dempsey KE, et al. Identification of bacteria on the surface of clinically infected and non-infected prosthetic hip joints removed during revision arthroplasties by 16S rRNA gene sequencing and by microbiological culture. Arthritis Res Ther 2007; 9(3):R46.

Fierer N, Hamady M, Lauber CL, Knight R. The influence of sex, handedness, and washing on the diversity of hand surface bacteria. Proc Natl Acad Sci U S A 2008; 105(46):17994–17999.

Fujimura KE, Slusher NA, Cabana MD, Lynch SV. Role of the gut microbiota in defining human health. Expert Rev Anti Infect Ther 2010; 8(4):435–454.

Garrett WS, Gordon JI, Glimcher LH. Homeostasis and inflammation in the intestine. Cell 2010; 140(6):859–870.

Hughes FJ, McNab R. Oral malodour—a review. Arch Oral Biol 2008; 53(Suppl 1):S1–S7.

Romagnani S. Coming back to a missing immune deviation as the main explanatory mechanism for the hygiene hypothesis. J Allergy Clin Immunol 2007; 119(6):1511–1513.

Taylor D, et al. Characterization of the microflora of the human axilla. Int J Cosmet Sci 2003; 25(3):137–145.

Thomas JG, Nakaishi LA. Managing the complexity of a dynamic biofilm. J Am Dent Assoc 2006; 137(Suppl):10S–15S.

Tiihonen K, Ouwehand AC, Rautonen N. Human intestinal microbiota and healthy ageing. Ageing Res Rev 2010; 9(2):107–116.

Turnbaugh PJ, et al. The human microbiome project. Nature 2007; 449(7164):804–810.

CHAPTER 21

Canny GO, McCormick BA. Bacteria in the intestine, helpful residents or enemies from within? Infect Immun 2008; 76(8):3360–3373.

Gotoh Y, et al. Two-component signal transduction as potential drug targets in pathogenic bacteria. Curr Opin Microbiol 2010; 13(2):232–239.

Otto M. Bacterial sensing of antimicrobial peptides. Contrib Microbiol 2009; 16: 136–149.

Shivaji S, Prakash JS. How do bacteria sense and respond to low temperature? Arch Microbiol 2010; 192(2):85–95.

CHAPTER 22

Capasso L. Bacteria in two-millennia-old cheese, and related epizoonoses in Roman populations. J Infect 2002; 45(2):122–127.

Caufield PW. Tracking human migration patterns through the oral bacterial flora. Clin Microbiol Infect 2009; 15(Suppl 1):37–39.

Cohn SK Jr, Weave LT. The Black Death and AIDS: CCR5Δ32 in genetics and history. QJM 2006; 99(8):497–503.

Monot M, et al. On the origin of leprosy. Science 2005; 308(5724):1040–1042.

Monot M, et al. Comparative genomic and phylogeographic analysis of *Mycobacterium leprae*. Nat Genet 2009; 41(12):1282–1289.

Morabia A. Epidemic and population patterns in the Chinese Empire (243 B.C.E. to 1911C.E.): quantitative analysis of a unique but neglected epidemic catalogue. Epidemiol Infect 2009; 137(10):1361–1368.

Wirth T, et al. Origin, spread and demography of the *Mycobacterium tuberculosis* complex. PLoS Pathog 2008; 4:e1000160

Yamaoka Y. *Helicobacter pylori* typing as a tool for tracking human migration. Clin Microbiol Infect 2009; 15(9):829–834.

CHAPTER 23

Bowey K, Neufeld RJ. Systemic and mucosal delivery of drugs within polymeric microparticles produced by spray drying. BioDrugs 2010; 24(6):359–377.

Caesar R, FÅk F, Bäckhed F. Effects of gut microbiota on obesity and atherosclerosis via modulation of inflammation and lipid metabolism. J Intern Med 2010; 268(4): 320–328.

Canfield DE, Glazer AN, Falkowski PG. The evolution and future of Earth's nitrogen cycle. Science 2010; 330(6001):192–196.

Collins SM, Bercik P. The relationship between intestinal microbiota and the central nervous system in normal gastrointestinal function and disease. Gastroenterology 2009; 136(6):2003–2014.

Jick H, Hagberg KW. Measles in the United Kingdom 1990–2008 and the effectiveness of measles vaccines. Vaccine 2010; 28(29):4588–4592.

Keshavarz T, Roy I. Polyhydroxyalkanoates: bioplastics with a green agenda. Curr Opin Microbiol 2010; 13(3):321–326.

Lidstrom ME, Konopka MC. The role of physiological heterogeneity in microbial population behavior. Nat Chem Biol 2010; 6(10):705–712.

Liu X, et al. Bioprospecting microbial natural product libraries from the marine environment for drug discovery. J Antibiot (Tokyo) 2010; 63(8):415–422.

Sekirov I, Russell SL, Antunes LC, Finlay BB. Gut microbiota in health and disease. Physiol Rev 2010; 90(3):859–904.

Stephens E, et al. Future prospects of microalgal biofuel production systems. Trends Plant Sci 2010; 15(10):554–564.

Tralau-Stewart CJ, Wyatt CA, Kleyn DE, Ayad A. Drug discovery: new models for industry-academic partnerships. Drug Discov Today 2009; 14(1–2):95–101.

CHAPTER 24

Chakraborty C, Mungantiwar AA. Human insulin genome sequence map, biochemical structure of insulin for recombinant DNA insulin. Mini Rev Med Chem 2003; 3(5):375–385.

Dellomonaco C, Fava F, Gonzalez R. The path to next generation biofuels: successes and challenges in the era of synthetic biology. Microb Cell Fact 2010; 9:3.

Gibson DG, et al. Creation of a bacterial cell controlled by a chemically synthesized genome. Science 2010; 329(5987):52–56.

Hohsaka T, Sisido M. Incorporation of non-natural amino acids into proteins. Curr Opin Chem Biol 2002; 6(6):809–815.

Salis H, Tamsir A, Voigt C. Engineering bacterial signals and sensors. Contrib Microbiol 2009; 16:194–225.

Schmidt FR. Recombinant expression systems in the pharmaceutical industry. Appl Microbiol Biotechnol 2004; 65(4):363–372.

Stockley PG, Twarock R. The physics of virus assembly. Phys Biol 2010; 7(4):040301.

Yadav VG, Stephanopoulos G. Reevaluating synthesis by biology. Curr Opin Microbiol 2010; 13(3):371–376.

CHAPTER 25

Horneck G, Klaus DM, Mancinelli RL. Space microbiology. Microbiol Mol Biol Rev 2010; 74(1):121–156.

La Duc MT, et al. Isolation and characterization of bacteria capable of tolerating the extreme conditions of clean room environments. Appl Environ Microbiol 2007; 73(8):2600–2611.

Moissl C, et al. Molecular bacterial community analysis of clean rooms where spacecraft are assembled. FEMS Microbiol Ecol 2007; 61(3):509–521.

Index of Bacterial Genera and Species

Bacteria: The Benign, the Bad, and the Beautiful, First Edition. Trudy M. Wassenaar.

Subject Index

Bacteria: The Benign, the Bad, and the Beautiful, First Edition. Trudy M. Wassenaar.